Form and Function in Developmental Evolution

This book represents a new effort to understand very old questions about biological form, function, and the relationships between them. The essays collected here reflect the diversity of approaches in evolutionary developmental biology (Evo Devo), including not only studies by prominent scientists whose research focuses on topics concerned with evolution and development, but also historically and conceptually oriented studies that place the scientific work within a larger framework and ask how it can be pushed further. Topics under discussion range from the use of theoretical and empirical biomechanics to understand the evolution of plant form, to detailed studies of the evolution of development and the role of developmental constraints on phenotypic variation. The result is a rich and interdisciplinary volume that will begin a wider conversation about the shape of Evo Devo as it matures as a field.

MANFRED D. LAUBICHLER is Professor of Theoretical Biology and History of Biology and Affiliated Professor of Philosophy at Arizona State University. He is co-editor with Gerd B. Müller of *Modeling Biology: Structures, Behaviors, Evolution* (2007).

JANE MAIENSCHEIN is Regents' Professor, President's Professor, Parents Association Professor, and Director of the Center for Biology and Society, Arizona State University. She is author of *Whose View of Life? Embryos, Cloning and Stem Cells* (2003, 2005) and co-editor with Manfred D. Laubichler of *From Embryology to Evo-Devo: A History of Developmental Evolution* (2007).

CAMBRIDGE STUDIES IN
PHILOSOPHY AND BIOLOGY

Recent Titles

Alfred I. Tauber *The Immune Self: Theory or Metaphor?*

Elliott Sober *From a Biological Point of View*

Robert Brandon *Concepts and Methods in Evolutionary Biology*

Peter Godfrey-Smith *Complexity and the Function of Mind in Nature*

William A. Rottschaefer *The Biology and Psychology of Moral Agency*

Sahotra Sarkar *Genetics and Reductionism*

Jean Gayon *Darwinism's Struggle for Survival*

Jane Maienschein and Michael Ruse (eds.) *Biology and the Foundation of Ethics*

Jack Wilson *Biological Individuality*

Richard Creath and Jane Maienschein (eds.) *Biology and Epistemology*

Alexander Rosenberg *Darwinism in Philosophy, Social Science, and Policy*

Peter Beurton, Raphael Falk and Hans-Jörg Rheinberger (eds.) *The Concept of the Gene in Development and Evolution*

David Hull *Science and Selection*

James G. Lennox *Aristotle's Philosophy of Biology*

Marc Ereshefsky *The Poverty of the Linnaean Hierarchy*

Kim Sterelny *The Evolution of Agency and Other Essays*

William S. Cooper *The Evolution of Reason*

Peter McLaughlin *What Functions Explain*

Steven Hecht Orzack and Elliot Sober (eds.)
Adaptationism and Optimality

Bryan G. Norton *Searching for Sustainability*

Sandra D. Mitchell *Biological Complexity and Integrative Pluralism*

Greg Cooper *The Science of the Struggle for Existence*

Joseph LaPorte *Natural Kinds and Conceptual Change*

Jason Scott Robert *Embryology, Epigenesis, and Evolution*

William F. Harms *Information and Meaning in Evolutionary Processes*

Marcel Weber *Philosophy of Experimental Biology*

Markku Oksanen and Juhani Pietorinen *Philosophy and Biodiversity*

Richard Burian *The Epistemology of Development, Evolution, and Genetics*

Ron Amundson *The Changing Role of the Embryo in Evolutionary Thought*

Sahotra Sarkar *Biodiversity and Environmental Philosophy*

Neven Sesardic *Making Sense of Heritability*

William Bechtel *Discovering Cell Mechanisms*

Giovanni Boniolo and Gabriele De Anna (eds.) *Evolutionary Ethics and Contemporary Biology*

Justin E.H. Smith (ed.) *The Problem of Animal Generation in Early Modern Philosophy*

Lindley Darden *Reasoning in Biological Discoveries*

Derek Turner *Making Prehistory*

Elizabeth A. Lloyd *Science, Politics, and Evolution*

Form and Function in Developmental Evolution

Edited by

MANFRED D. LAUBICHLER

Arizona State University

JANE MAIENSCHEIN

Arizona State University

CAMBRIDGE UNIVERSITY PRESS
Cambridge, New York, Melbourne, Madrid, Cape Town, Singapore,
São Paulo, Delhi, Dubai, Tokyo

Cambridge University Press
The Edinburgh Building, Cambridge CB2 8RU, UK

Published in the United States of America by Cambridge University Press, New York

www.cambridge.org
Information on this title: www.cambridge.org/9780521872683

First published 2009

A catalogue record for this publication is available from the British Library

Library of Congress Cataloguing in Publication data
Form and function in developmental evolution / edited by Manfred D.
Laubichler, Jane Maienschein.
p. cm. – (Cambridge studies in philosophy and biology)
Includes bibliographical references and index.
ISBN 978-0-521-87268-3 (hardback)
1. Developmental biology. 2. Evolution (Biology) 3. Morphology.
I. Laubichler, Manfred Dietrich II. Maienschein, Jane.
III. Title. IV. Series.
QH366.2.F675 2008
571.8–dc22
2009000110

ISBN 978-0-521-87268-3 Hardback

Transferred to digital printing 2010

Contents

Figures

Tables

Contributors

Paul M. Brakefield is Professor of Evolutionary Biology at Leiden University. He is the author of numerous studies integrating experimental laboratory studies with fieldwork focusing on the patterns of butterfly wings.

Andrew L. Hamilton is Assistant Professor at the Center for Biology and Society and the Center for Social Dynamics and Complexity in the School of Life Sciences at Arizona State University. He is also Affiliated Professor in the Philosophy Department at ASU, and Assistant Director of History and Philosophy of Taxonomy at the International Institute for Species Exploration. He has published widely on conceptual issues in biology, including selection at higher taxa and the evolution of cooperative behavior in social insects.

Manfred D. Laubichler is Professor of Theoretical Biology and History of Biology and Affiliated Professor of Philosophy at the School of Life Sciences and Centers for Biology and Society and Social Dynamics and Complexity at Arizona State University. He is the coeditor of *From Embryology to Evo-Devo* (2007), *Modeling Biology* (2007), and *Der Hochsitz des Wissens* (2006), and an associate editor of *Endothelial Biomedicine* (Cambridge University Press, 2007). He is also an associate editor of the *Journal of Experimental Zoology, Part B: Molecular and Developmental Evolution* and of *Biological Theory*.

Jane Maienschein is Regents' Professor, President's Professor, and Parents Association Professor at Arizona State University, where she also directs the Center for Biology and Society. She is (co)editor of a dozen volumes and author of, most recently, *Whose View of Life?*

Embryos, Cloning, and Stem Cells (2003, 2005). She serves as President of the History of Science Society and was the former editor of the *Journal of the History of Biology*.

Karl J. Niklas is the Liberty Hyde Bailey Professor of Plant Biology at Cornell University, President of the American Botanical Society, and a former editor of the *American Journal of Botany*. He is the author of *Plant Biomechanics* (1992), *Plant Allometry* (1994), and *The Evolutionary Biology of Plants* (1997).

Elizabeth C. Raff is Professor of Biology and Chair of the Department of Biology at Indiana University. She is the author of numerous publications in molecular and developmental biology.

Rudolf A. Raff is Distinguished Professor and James H. Rudy Professor of Biology at Indiana University. He is the author of *Embryos, Genes, and Evolution: The Developmental Genetic Basis of Evolutionary Change*, with T. Kaufman (1983), and of *The Shape of Life: Genes, Development, and the Evolution of Animal Form* (1996). He is also the editor of the journal *Evolution and Development.*

Richard A. Richards is Associate Professor of Philosophy at the University of Alabama. He has published widely in the area of philosophy of biology. In addition, as a former dancer, he is involved in various artistic projects and has been artist-in-residence at Yosemite National Park.

Roger Sansom is Assistant Professor of Philosophy at Texas A&M University and the coeditor of *Integrating Evolution and Development: From Theory to Practice* (2007).

Günter P. Wagner is the Alison Richard Professor of Ecology and Evolutionary Biology and chair of the Department of Ecology and Evolutionary Biology at Yale University. He is the editor in chief of the *Journal of Experimental Zoology*, and editor of *The Character Concept in Evolutionary Biology* (2001) and *Modularity in Development and Evolution* (2005).

Peter C. Wainwright is Professor of Evolutionary Biology and Ecology at the University of California, Davis. He is a Fellow of the American Association for the Advancement of Science and the author of numerous publications in functional morphology and biomechanics of vertebrates, ecological morphology of fish, and phylogenetics.

Journal abbreviations

Acta Anat	*Acta Anatomica*
Am Nat	*The American Naturalist*
Am Zool	*American Zoologist*
Anim Behav	*Animal Behaviour*
Arch Biol (Bruxelles)	*Archives de biologie (Bruxelles)*
Behav Ecol Sociobiol	*Behavioural Ecology and Sociobiology*
Biol Bull	*The Biological Bulletin*
Biol J Linn Soc Lond	*Biological Journal of the Linnean Society, Linnean Society of London*
Biol Rev Camb Philos Soc	*Biological Reviews of the Cambridge Philosophical Society*
Bull Soc Hist Nat Toulouse	*Bulletin de la Société de l'histoire naturelle (Toulouse)*
Curr Biol	*Current Biology*
Curr Opin Genet Dev	*Current Opinion in Genetics & Development*
Curr Opin Neurobiol	*Current Opinion in Neurobiology*
Dev Biol	*Developmental Biology*
Evol Biol	*Evolutionary Biology*
Evol Dev	*Evolution & Development*
Genome Res	*Genome Research*
Harvey Lect	*Harvey Lectures*
Int J Dev Biol	*International Journal of Developmental Biology*
J Biosci	*Journal of Biosciences*
J Embryol Exp Morphol	*Journal of Embryology and Experimental Morphology*
J Evol Biol	*Journal of Evolutionary Biology*

J Exp Mar Bio Ecol	*Journal of Experimental Marine Biology and Ecology*
J Exp Zool	*Journal of Experimental Zoology*
J Exp Zoolog B Mol Dev Evol	*Journal of Experimental Zoology. Part B. Molecular and Developmental Evolution*
J Herpetology	*Journal of Herpetology*
J Med Genet	*Journal of Medical Genetics*
J Morphol	*Journal of Morphology*
J Paleontol	*Journal of Paleontology*
J Vert Paleont	*Journal of Vertebrate Paleontology*
Mem Calif Acad Sci	*Memoirs of the California Academy of Science*
Nat Rev Genet	*Nature Reviews Genetics*
Novartis Found Symp	*Novartis Foundation Symposium*
Pediatr Nephrol	*Pediatric Nephrology*
PNAS	*Proceedings of the National Academy of Sciences*
Proc Biol Sci	*Proceedings. Biological Sciences*
Proc Natl Acad Sci USA	*Proceedings of the National Academy of Sciences USA*
Proc R Soc Lond B	*Proceedings of the Royal Society London B: Biological Sciences*
PSA	*Proceedings of the Biennial Meeting of the Philosophy of Science Association*
Q Rev Biol	*Quarterly Review of Biology*
Syst Biol	*Systematic Biology*
Syst Zool	*Systematic Zoology*
Theor Popul Biol	*Theoretical Population Biology*
Trends Ecol Evol	*Trends in Ecology & Evolution*
Trends Genet	*Trends in Genetics*
Trop Zoology	*Tropical Zoology*
Z Anat Entwicklungsgesch	*Zeitschrift für Anatomie und Entwicklungsgeschichte*
Zool J Linn Soc	*Zoological Journal of the Linnean Society*
Zool Jahrb Abt Anat Ontogenie Tiere	*Zoologische Jahrbücher. Abteilung für Anatomie und Ontogenie der Tiere*

1

Introduction

ANDREW L. HAMILTON, MANFRED D. LAUBICHLER,
AND JANE MAIENSCHEIN

When philosopher of biology Michael Ruse moved from Guelph, Ontario, to Florida State University, a very good thing happened for our understanding of the conceptual, historical, and philosophical foundations of biology. Ruse became the William H. and Lucyle T. Werkmeister Professor, which brought with it an endowment that allowed him to organize conferences on a regular basis. As usual, Michael Ruse lost no time, and the resulting series of conferences has brought together biologists, historians, and philosophers in lively discussion of a number of important topics. A favorite image of those events is that of Ernst Mayr, sitting in the boat during a swamp tour on a drowsy Florida afternoon. The aging but ever-intense Mayr seemed to be dozing, when suddenly he pointed and declared a noteworthy bird, then another, and another. That led to a discussion of whether biodiversity is declining; then to philosophical questions about how we count diversity; and finally to ethical and policy questions of why we care. Other conferences have led to debates about science and religion. And so on.

In 2005, a group of leading biologists joined philosophers and historians for four days of thinking about form and function. For this meeting, Ruse followed his usual approach. He provided the general theme, brought together a mix of enthusiastic scholars, and waited to see what happened. In this case, it was something very interesting.

While some of the papers looked at more traditional questions related to form or function, or even the two together, most asked questions about form and function in light of the (still) new emphasis on developmental evolution. They tied together what would typically have been a broad range of quite different approaches by people who would not ordinarily have been talking to each other. The (unintended) unifying theme of this conference was how form and function relate to larger issues of Evo Devo – that is, by thinking about how development informs considerations of evolution, traditional form and function questions are transformed in ways that bring them together as reflections on this new organizing theme. What we offer here are not conference proceedings, but papers that grew out of subsequent discussions inspired by that conference.

We thank Michael Ruse for stimulating the discussion, the William H. and Lucyle T. Werkmeister Endowment Fund for making such an intellectual synthesis possible, and Hilary Gaskin at Cambridge University Press for her patience as we have worked out the logistics. Our work on this volume was supported by NSF SES 0645729 to MDL and NSF SES 0623176 to MDL and JM. We offer this set of papers as an invitation to others from diverse disciplines to join the discussion. Far from addressing all possible questions or providing a summary for a mature field, this volume serves as an invitation that provides a rich collection of papers that point toward new questions and new directions for research in problems of form and function.

The papers collected here each represent a major research emphasis connected to the problem of the relationship of form and function within the diversity of approaches that make up twenty-first-century Evo Devo. Manfred Laubichler first provides a historical and conceptual analysis of the treatments of form and function within the framework of Evo Devo. He shows that many of the traditional issues connected to the relationship of form and function predate genuine Evo Devo questions, and that several of the late twentieth-century origins of Evo Devo have been focused explicitly on the relationship of form and function (Laubichler and Maienschein 2007a; Laubichler and Maienschein 2007b). Based on his analysis of the research programs, central questions, and unifying concepts of present-day Evo Devo, and current work on the developmental evolution of social behavior, Laubichler then argues that a mechanistic account of developmental evolution offers a solution to the age-old problem of integrating form and function (Laubichler 2005; Laubichler 2007a; Laubichler 2007b; Laubichler and Gadau in press; Laubichler and Müller 2007). Such a mechanistic framework of developmental evolution is based on an understanding of the general principles and molecular details of developmental systems governing phenotypic characters, and the identification of the causal connections between variation in those developmental systems and the observed patterns of phenotypic variation. It also suggests concrete evolutionary scenarios of how underlying developmental changes govern evolutionary transformations of phenotypes and experimental tests to uncover the selective forces driving these evolutionary transformations. This way, Laubichler proposes, the mechanistic framework of developmental evolution unites the perspectives of form and function that have so often been associated with separate explanatory frameworks and subdisciplines.

Karl J. Niklas's research program has led to remarkable insights into the general principles of plant morphology and their evolutionary origins, and it also suggests how the lessons from theoretical morphology, biomechanics, and functional analysis can be combined to arrive at a detailed understanding of plant evolution (Niklas 1994; Niklas 1997). Furthermore, his approach incorporates the developmental principles of morphogenesis and, wherever the data are available, plant developmental genetics to connect his phenotypic analysis of form and function to the larger explanatory framework of Evo Devo. Niklas's approach further demonstrates that the notion of constraint, a main focus of the first wave of Evo Devo proposals, does not just impose limits on variation, but rather is a constitutive element of any morphogenetic process that both enables and limits phenotypic possibilities.

Rudy and Elizabeth Raff's research program focuses on the relationship between developmental and evolutionary processes and, in particular, on several foundational questions of ontogeny and phylogeny, such as the origin of different modes of development (Raff 1996; Raff and Raff 2000; Raff and Raff 2007; Raff *et al.* 2003). To this end, they have adapted a unique model system: a pair of sister taxa of sea urchins with different modes of development – one direct developer and one with larval development. Taking advantage of all the tools of molecular biology, together with some more traditional approaches such as species hybridization techniques, they are now able to uncover the molecular basis of different developmental systems. This work has led to a re-evaluation of some of the most entrenched assumptions about the relationship between ontogeny and phylogeny. Even though Haeckel's phylogenetic law has in its narrow sense been disproven for a long time, the fundamental idea that earlier stages in development are more conserved than later ones continues to be widely held (Churchill 2007). The data that Rudy and Elizabeth Raff have collected over the years show that fundamental larval and developmental features can change relatively fast during evolution. Their work thus connects micro- with macroevolutionary perspectives, as well as comparative embryology (with its emphasis on form) with the ecological (functional) conditions that drive the evolution of different larval and developmental modes.

Peter Wainwright's main concern is the problem of innovation and how it relates to morphological diversity (Ferry-Graham *et al.* 2001; Hulsey and Wainwright 2002; Wainwright 2002). Several models of evolution relate rapid speciation events and their corresponding degrees of phenotypic variation to the emergence of key innovations that enable a group of species to conquer new territories or exploit new resources (Schluter 2000). These adaptive radiations are a prime example of how functional advantages drive evolutionary change; but as many of these functional benefits are a consequence of specific structures (form), it is difficult to disentangle form and function in these cases. The basic premise of evolutionary functional morphology is that form enables function and that function, measured by fitness, feeds back on the further refinement of forms or structures. In these cases, form and function are thus seen as complementary. But what exactly is a key innovation and how do we detect it in relevant datasets? This is not a trivial question. Simply assuming that whenever a group with a high number of species shares a common character (synapomorphy) this character has to be a key innovation is a clear case of tautological reasoning. Similarly, assuming that a specific character that is sufficiently different from others dramatically changes the evolutionary dynamics of the morphospace for all the species that share it also needs to be tested against objective measurements. Wainwright presents such a test in the form of a newly developed algorithm that compares and estimates rates of morphological evolution. Such measures of the rate of trait evolution control for confounding effects, such as time or shared evolutionary history, and therefore allow distinguishing adaptive radiations from normal or baseline rates of morphological evolution.

Paul Brakefield also focuses on the integration of form and function – in his case, the specific patterns of variation of butterfly eyespots (Beldade and Brakefield 2002; Beldade and Brakefield 2003; Beldade *et al.* 2005; Beldade *et al.* 2008; Brakefield 2001; Brakefield 2006; Brakefield 2007; Brakefield and French 2005; Brakefield and Roskam 2006; Brakefield *et al.* 1996; Brakefield *et al.* 2007). The butterfly eyespots are particularly interesting characters for addressing the relationship of form and function; not only do they have clear adaptive functions, but we also know quite a lot about the developmental mechanisms causing these phenotypes. Eyespots are thus prime examples of an integrative perspective on Evo Devo, one that combines form and function, or the internal developmental and external ecological processes of phenotypic evolution. Brakefield's particular combination of approaches – artificial selection experiments in the wild and laboratory analysis of developmental systems – allows him to address the degree to which the mechanisms of development facilitate and constrain phenotypic variation, and also to identify potential targets of selection among the regulatory elements of the developmental system. Discovering the allometric relationships underlying phenotypic transformations of the eyespot reveals the mechanistic underpinnings of how developmental processes are producing variation as the raw material for natural selection to act on. Brakefield's model system is thus a perfect example of how a combination of different approaches can lead to a mechanistic model of phenotypic evolution that combines the perspectives of both form and function and of evolutionary and developmental biology.

Günter Wagner presents a case study that focuses on a conflict of evidence between different research traditions and types of evidence, paleontology and developmental biology, in explanations of avian digit identity (Galis *et al.* 2002; Galis *et al.* 2005; Stopper and Wagner 2005; Vargas and Fallon 2005a; Vargas and Fallon 2005b; Wagner and Gauthier 1999; Wagner and Müller 2002). This case is instructive in many ways. It highlights the different explanatory frameworks of comparative anatomy and paleontology and experimental developmental biology. But it also points to the fact that in many cases the resolution of such conflicts does not lie in asking which of the conflicting interpretations has the stronger support – which would be akin to the old question whether form or function is the prime cause in explanations of biological phenomena – but rather in a conceptual innovation, in this case the frame-shift hypothesis, that enables us to integrate different types of evidence within one inclusive mechanistic model. Wagner's case study is therefore a perfect example of how Evo Devo can provide conceptual resolution and synthesis to some of the traditional antagonisms and conflicts within biological theories.

The papers by Roger Sansom and Richard Richards represent some of the recent philosophical work in response to the new Evo Devo orientation within biology (Amundson 2005; Sansom and Brandon 2007; Wimsatt 2007). In general, from its early inception in the 1970s, Evo Devo as a field has sought the contact of both historians and philosophers of biology. This has arguably been a most productive relationship. One of the reasons why scientists have found this contact profitable lies precisely in the integrative and interdisciplinary nature of

Evo Devo, which benefits from the kind of conceptual analyses that we see in both papers. The difficulty of establishing character transformation series (Richards) and the question of constraints (Sansom) are both central problems of Evo Devo that are connected to the problem of the relationship of form and function. Both are also in many ways foundational to the empirical studies represented in this volume.

Richards analyzes one of the central issues connected to the relationship of phylogeny and morphological evolution/character transformation. He argues that the question of whether functional considerations are a legitimate part of phylogenetic inference is ultimately an empirical one, but by delineating the various theoretical assumptions behind the two competing approaches – one privileging form, the other function – he has contributed in significant ways to the eventual resolution of this debate.

Similarly, Roger Sansom's paper presents a conceptual analysis of the notion of constraint and its roles in explanations of phenotypic evolution. As in Richard's case, Sansom does not "solve" the problem, but he provides a useful road map for future discussions, both within the philosophy of biology and within Evo Devo itself.

In a final paper, Andrew Hamilton picks up the theme of mechanistic explanation from the volume's first chapter, asking how such explanations work, what challenges arise from this approach, and how mechanistic thinking might integrate form and function on the one hand and evolutionary biology and developmental biology on the other. The paper has two central themes. The first is a discussion of the ways in which a mechanistic Evo Devo that focuses heavily on gene regulatory networks is and is not reductionistic. After arguing that mechanistic thinking leads Evo Devo toward a responsible variety of reductionism, Hamilton moves on to discuss a specific new challenge for mechanistic Evo Devo: how to understand and explain what he calls "levels of development." Hamilton's concern with levels of development grows out of asking what happens when an epistemic commitment to mechanistic thinking is combined with the ontological commitment that colonies of social insects are "superorganisms" in something more than a metaphorical sense. The challenge of the paper is for researchers to find ways to ask and answer what it means to give a mechanistic explanation of development at the colony level that is informed by what we know about development at the organismal level, as well as how the two levels inform each other's evolution.

This collection contains an embarrassment of riches. The biological studies included here should serve as a starting point for further and deeper conversations about form and function and what they mean against the backdrop of Evo Devo as a biological project. Evo Devo, of course, is as much a conceptual project as an experimental one, and we hope that the context provided here by the historical and philosophical chapters stimulates more discussion about the scope, goals, methods, assumptions, and aims of Evo Devo. These pieces taken as a whole show something of the exciting intellectual pursuit that lies ahead as Evo Devo serves as a framework for addressing long-standing issues like the relationship between form and function. Evo Devo is still new, but it is also maturing. The structure that is coming into view holds promise, and there is much work to be done in understanding how form, function, evolution, and development go together to form a nuanced picture of the biological world.

REFERENCES

Amundson, R. (2005). *The Changing Role of the Embryo in Evolutionary Thought: Roots of Evo-Devo.* Cambridge and New York: Cambridge University Press.

Beldade, P. and Brakefield, P.M. (2002). The genetics and evo-devo of butterfly wing patterns. *Nat Rev Genet* 3, 442–52.

(2003). Concerted evolution and developmental integration in modular butterfly wing patterns. *Evol Dev* 5, 169–79.

Beldade, P., Brakefield, P.M., and Long, A.D. (2005). Generating phenotypic variation: prospects from "evo-devo" research on Bicyclus anynana wing patterns. *Evol Dev* 7, 101–7.

Beldade, P., French, V., and Brakefield, P.M. (2008). Developmental and genetic mechanisms for evolutionary diversification of serial repeats: eyespot size in Bicyclus anynana butterflies. *J Exp Zoology B Mol Dev Evol*, 310B, 191–201.

Brakefield, P.M. (2001). Structure of a character and the evolution of butterfly eyespot patterns. *J Exp Zool* 291, 93–104.

(2006). Evo-devo and constraints on selection. *Trends Ecol Evol* 21, 362–8.

(2007). Butterfly eyespot patterns and how evolutionary tinkering yields diversity. *Novartis Found Symp* 284, 90–101; discussion 101–15.

Brakefield, P.M. and French, V. (2005). Evolutionary developmental biology: how and why to spot fly wings. *Nature* 433, 466–7.

Brakefield, P.M., Gates, J., Keys, D., Kesbeke, F., Wijngaarden, P.J., Monteiro, A., French, V., and Carroll, S.B. (1996). Development, plasticity and evolution of butterfly eyespot patterns. *Nature* 384, 236–42.

Brakefield, P.M., Pijpe, J., and Zwaan, B.J. (2007). Developmental plasticity and acclimation both contribute to adaptive responses to alternating seasons of plenty and of stress in Bicyclus butterflies. *J Biosci* 32, 465–75.

Brakefield, P.M. and Roskam, J.C. (2006). Exploring evolutionary constraints is a task for an integrative evolutionary biology. *Am Nat* 168, Suppl. 6, S4–13.

Churchill, F.B. (2007). Living with the biogenetic law: a reappraisal. In M.D. Laubichler and J. Maienschein (eds.), *From Embryology to Evo Devo: A History of Developmental Evolution*. Cambridge, MA: MIT Press, pp. 37–82.

Ferry-Graham, L.A., Wainwright, P.C., and Bellwood, D.R. (2001). Prey capture in long-jawed butterflyfishes (Chaetodontidae): the functional basis of novel feeding habits. *J Exp Mar Bio Ecol* 256, 167–84.

Galis, F., Kundrat, M., and Metz, J.A. (2005). Hox genes, digit identities and the theropod/bird transition. *J Exp Zoolog B Mol Dev Evol* 304, 198–205.

Galis, F., van Alphen, J.J., and Metz, J.A. (2002). Digit reduction: via repatterning or developmental arrest? *Evol Dev* 4, 249–51.

Hulsey, C.D. and Wainwright, P.C. (2002). Projecting mechanics into morphospace: disparity in the feeding system of labrid fishes. *Proc Biol Sci* 269, 317–26.

Laubichler, M.D. (2005). Evolutionäre Entwicklungsbiologie. In U. Krohs and G. Toepfer (eds.), *Einführung in die Philosophie der Biologie*. Frankfurt/Main: Suhrkamp, pp. 322–37.

(2007a). Evolutionary developmental biology. In D. Hull and M. Ruse (eds.), *Cambridge Companion to the Philosophy of Biology*. Cambridge University Press, pp. 342–600.

(2007b). The regulatory genome: Eric Davidson at 70. *Bioessays* 29, 937–9.

Laubichler, M.D. and Gadau, J. (in press). Social insects as models for evo devo. In J. Gadau and J. Fewell (eds.), *Organization of Insect Societies: From Genomes to Socio-Complexity*. Cambridge, MA: Harvard University Press.

Laubichler, M.D. and Maienschein, J. (2007a). Embryos, cells, genes, and organisms: a few reflections on the history of evolutionary developmental biology. In R. Brandon and R. Sansom (eds.), *Integrating Evolution and Development: From Theory to Practice*. Cambridge, MA: MIT Press, pp. 1–24.

(2007b). *From Embryology to Evo-Devo: A History of Developmental Evolution*. Cambridge, MA: MIT Press.

Laubichler, M. and Müller, G.B. (2007). *Modeling Biology: Structures, Behavior, Evolution*. Cambridge, MA: MIT Press.

Niklas, K.J. (1994). *Plant Allometry: The Scaling of Form and Process*. University of Chicago Press.

(1997). *The Evolutionary Biology of Plants*. University of Chicago Press.

Raff, E.C., Popodi, E.M., Kauffman, J.S., Sly, B.J., Turner, F.R., Morris, V.B., and Raff, R.A. (2003). Regulatory punctuated equilibrium and convergence in the evolution of developmental pathways in direct-developing sea urchins. *Evol Dev* 5, 478–93.

Raff, E.C. and Raff, R.A. (2000). Dissociability, modularity, evolvability. *Evol Dev* 2, 235–7.

Raff, R.A. (1996). *The Shape of Life: Genes, Development, and the Evolution of Animal Form*. University of Chicago Press.

Raff, R.A. and Raff, E.C. (2007). Tinkering: new embryos from old – rapidly and cheaply. *Novartis Found Symp* 284, 35–45; discussion 45–54, 110–15.

Sansom, R. and Brandon, R.N. (2007). *Integrating Evolution and Development: From Theory to Practice*. Cambridge, MA: MIT Press.

Schluter, D. (2000). *The Ecology of Adaptive Radiation.* Oxford University Press.

Stopper, G.F. and Wagner, G.P. (2005). Of chicken wings and frog legs: a smorgasbord of evolutionary variation in mechanisms of tetrapod limb development. *Dev Biol* 288, 21–39.

Vargas, A.O. and Fallon, J.F. (2005a). Birds have dinosaur wings: the molecular evidence. *J Exp Zoolog B Mol Dev Evol* 304, 86–90.

(2005b). The digits of the wing of birds are 1, 2, and 3. A review. *J Exp Zoolog B Mol Dev Evol* 304, 206–19.

Wagner, G.P. and Gauthier, J.A. (1999). 1,2,3 = 2,3,4: a solution to the problem of the homology of the digits in the avian hand. *Proc Natl Acad Sci USA* 96, 5111–16.

Wagner, G.P. and Müller, G.B. (2002). Evolutionary innovations overcome ancestral constraints: a re-examination of character evolution in male sepsid flies (Diptera: Sepsidae). *Evol Dev* 4, 1–6; discussion 7–8.

Wainwright, P.C. (2002). The evolution of feeding motor patterns in vertebrates. *Curr Opin Neurobiol* 12, 691–5.

Wimsatt, W.C. (2007). *Re-engineering Philosophy for Limited Beings: Piecewise Approximations to Reality.* Cambridge, MA: Harvard University Press.

2

Form and function in Evo Devo: historical and conceptual reflections

MANFRED D. LAUBICHLER

The success of grand historical narratives and conceptual reflections rests, in no small part, on the selection of their central characters and features. If, as in our case here, the focus is on the development and integration of those ideas that have shaped our understanding of the living world, it is hardly possible to think of one pair of concepts more appropriate to structure a *longue durée* narrative than form and function. These linked concepts are uniquely suited to organizing the rich diversity and idiosyncratic developments within the history and philosophy of the life sciences in general, and within evolutionary developmental biology in particular.

The concepts of form and function, and everything connected to them, have been used as an organizing principle in several influential analyses of the development of the biological sciences. Their symbiotic and often dialectic relationship forms the backbone of such treatises as Ernst Mayr's *The Growth of Biological Thought*, William Coleman's *Biology in the Nineteenth Century*, Ernst Cassirer's *Theory of Knowledge* (specifically the section on the biological sciences), Edmund Beecher Wilson's *The Cell in Development and Heredity*, and, of course, Edward Stuart Russell's *Form and Function: A Contribution to the History of Animal Morphology* (Cassirer 1950; Coleman 1971; Mayr 1982; Russell 1916; Wilson 1925).

Focusing on the concepts of form and function rather than the numerous historical and empirical details of biological research allowed these scholars to capture what Russell called "typical attitudes" and their connected theoretical positions, supporting empirical evidence and the historical dynamics of scientific change. Such an approach is also useful for us, as we seek to localize the main theoretical positions of present-day evolutionary developmental biology, represented in this volume in chapters by some of the leading practitioners in the field, within both the larger conceptual framework of the biological sciences and their multi-layered dynamical history. *Form and Function in Developmental Evolution* thus represents a collective endeavor that highlights the connections between the historical development of ideas, the conceptual foundations of present-day Evo Devo and their empirical basis, and the philosophical problems connected with these issues.

This chapter places twenty-first-century Evo Devo within the historical trajectory represented by form and function in two ways: it will show (1) how many of the traditional problems of form and function are predecessors of genuine Evo Devo questions; and (2) how several of the multiple late twentieth-century origins of Evo Devo have been focused explicitly on the relationship of form and function. Many early proponents of twentieth-century Evo Devo emphasized the need to study form as a corrective to what they perceived to be an over-reliance on functional (fitness) or optimality (adaptation) type explanations in evolutionary biology (Amundson 2005; Arthur 2002; Bonner 1958; Gould 1977; Hall 1998; Laubichler and Maienschein 2007b; Love and Raff 2003). This chapter analyzes how the relationship between form and function is represented in the main research programs, research questions, and concepts of Evo Devo. This relationship highlights what is arguably the main challenge of Evo Devo: how to integrate different perspectives on form and function (notably those that emphasize developmental mechanisms and those that focus on evolutionary processes) within one consistent explanatory framework.

This last problem, as we will see throughout this volume, has a potentially viable solution in the form of a mechanistic theory of developmental evolution. Such a theory unites the traditionally separate conceptual perspectives of form and function within the parameters of a mechanistic understanding of the evolutionary consequences of developmental processes. A major part of this emerging theory focuses on the role of gene regulatory networks for our understanding of both differentiation during development and transformation during evolution (Davidson 2006a; Davidson and Erwin 2006). Gene regulatory networks explain how different genes are turned on and off in separate cells and tissues during development, while the known architecture of these networks, and specifically the fact they these are composed of subcircuits that show varying degrees of conservation, correlates with observed patterns of phenotypic evolution (Davidson 2006a; Davidson and Erwin 2006; Riedl 1975). Large systematic groups not only share a morphological bauplan, but they also possess the same conserved regulatory sub-circuits, which, in turn, explain the remarkable stability of these morphological features. Phenotypic evolution, the explanandum of Evo Devo, can then be understood in terms of changes within the regulatory machinery of development. A mechanistic theory of developmental evolution based on these gene regulatory networks thus integrates the different timescales of ontogeny and phylogeny.

This new theoretical solution to the age-old problem of bringing together perspectives of form and function has become possible after many of the molecular details of the mechanisms of developmental differentiation, especially the limited number of structural genes and the features of the regulatory circuits, including their high degrees of conservation, have become known on a comparative scale. The conceptual forerunners of this proposed modern integration of form and function, however, date back to the turn of the nineteenth century and the emergence of morphology, functional anatomy, and comparative anatomy and embryology. Let us then briefly turn to the nineteenth century and reconstruct those elements of the complex dynamics between form and function that are relevant to the topic under discussion here.

FORM AND FUNCTION IN NINETEENTH-CENTURY BIOLOGY: SETTING THE STAGE FOR TWENTY-FIRST-CENTURY EVO DEVO

The nineteenth century provides the conceptual backdrop to several of the developments now at the core of twenty-first-century Evo Devo. As the nineteenth century progressed, and biology slowly changed from a concept to a more mature science, ideas about form and function and their complex relationships were transformed in the context of several emerging research programs and theoretical innovations (Ballauff and Ungerer 1954; Bowler 1983; Bowler 1996; Bowler 2003; Bowler and Morus 2005; Coleman 1971; Gould 1977; Jahn 1998; Laubichler and Maienschein 2007b; Nordenskiöld 1967; Russell 1916; Sapp 2003).

The relationship between form and function was already at the heart of the emerging sciences of morphology and comparative anatomy at the turn of the nineteenth century (Appel 1987; Breidbach 2006; Breidbach and Ziche 2001; Breidbach *et al.* 2001; Goethe 1824; Goethe and Günther 1981; Nyhart 1995; Richards 2002; Roger and Williams 1997; Rudwick and Cuvier 1997). The conceptual innovation of these new approaches to the question of how organic forms arise was that they were no longer content with description or classification, but aimed to provide a causal explanation for both form and function that was more than a simple affirmation of design or the economy of nature. Even though Kant had declared that there will never be a "Newton of a blade of grass," Goethe (for morphology) and Cuvier (for comparative and functional anatomy) attempted to be just that. They both searched for laws and rules governing the observed patterns of organic forms. For Goethe, these laws were the principles of metamorphosis (Goethe 1824). His approach focused on internal or generative principles of metamorphosis that allowed him not only to explain all existing forms (of flowering plants), but also to deduce all those forms that could exist. In this account, the principles of form are primary and function follows form. Goethe's interest in generative principles reflected to no small degree his artistic inclinations, and was enthusiastically received by proponents of *Naturphilosophie* as well as by romantic artists (Richards 2002). It is, as yet, an open question to what degree Darwin was influenced by these discussions when he emphasized the internal conditions and embryological origins of phenotypic variation (Richards 1992).

Cuvier, on the other hand, was as pragmatic as Goethe was artistic, in his quest to discover the laws governing organic forms. He too emphasized internal principles, such as the correlation between parts, but he also clearly saw form as connected to its function. The various anatomical structures of a carnivore, for instance, are perfectly adapted to its way of life; whether these are teeth or claws, the structure of the digestive tract, or the sensory organs, they all derive their purpose from the specific behavior of a carnivore. These insights from functional anatomy received further confirmation from a comparative perspective that revealed their generality – that is, many different carnivores display similar correlations between their functional parts (Rudwick and Cuvier 1997). This combination of comparative and functional perspectives and their associated principles allowed Cuvier to pronounce that he could reconstruct any animal (including extinct) from knowing just one bone. Cuvier thus differed from Goethe in that the purpose or function of an anatomical structure was an essential part of its explanation.

In the wake of Goethe's and Cuvier's theoretical arguments, the sciences of form entered a period of growth and differentiation, with each camp attracting followers and further advancing its theoretical and empirical basis. Goethe had already emphasized transformation (metamorphosis) as an important part of explanations of form, and this dynamic perspective would soon dominate nineteenth-century discussions of anatomy and morphology. Embryological evidence was soon added to explanations of form and the developmental origin of anatomical structures became an important area of investigation, transforming itself from the largely speculative theories of the seventeenth and eighteenth centuries into a comparative descriptive and later experimental science (Laubichler and Maienschein 2004; Laubichler and Maienschein 2007a). The embryological approach dramatically changed explanations of form. These would now be framed in the context of their predecessors, or anlagen, and the mechanisms that transformed those into adult structures. These mechanisms included physical forces (of bending and folding tissues), chemical gradients, and, later, internal factors of differentiation (first seen as either nuclear or cytoplasmic factors and later as nuclear genes).

The comparative embryological orientation of the sciences of form also paved the way for the theory of evolution. Comparing different developmental sequences suggested that recapitulation of embryological stages and subsequent terminal addition of new features could be an explanation of the observed patterns of phenotypic variation. In its first version, the theory of recapitulation (proposed by Meckel and Serres) was rejected by Karl-Ernst von Baer, who argued for the independence of each developmental trajectory (Gould 1977). But von Baer also observed that the early stages of development are more similar than later stages. Independent of the final verdict on recapitulation, the emphasis on embryological data as part of explanations of adult forms proved to be very productive indeed, and became a major methodological and causal factor in reconstructions of phylogeny after Darwin's convincing argument for evolution (Gould 1977; Laubichler and Maienschein 2007a).

A lot is known about the many sources that contributed to Darwin's eventual formulation of the theory of descent. These studies have made it abundantly clear that embryology played a major role in Darwin's arguments, both as evidence for the relatedness of different taxa, and as a mechanism that helps us to understand the patterns of existing phenotypic variation (the laws governing organic forms), as well the origin of new variants (the big challenge for his theory). But Darwin's focus on the survival and reproductive success of individuals within populations as the causal agent of evolution also emphasized functional considerations. Structures exist because they contribute to the fitness of individuals who have them. This functional part of Darwin's theory would become a dominant feature of the mid-twentieth-century Modern Synthesis. For Darwin himself, both the principles and laws of form – especially those connected with development – and the functional principles of survival were an integral part of explanations of evolution (Kohn and Kottler 1985; Richards 1992).

Functional considerations, however, were not particularly useful for reconstructing phylogenies – the major research emphasis of evolutionary morphology and especially the Haeckel–Gegenbaur school. For this purpose comparative anatomy, especially the establishment of homologies, and comparative embryology proved initially to be more productive. An embryological criterion was seen as the most desirable explanation for homologies. This emphasis on form had its own difficulties, however. Without a better mechanistic understanding of development, many questions about the identity of embryological structures could not be decided, and this lack of clear evidential criteria was also the main reason why the program of evolutionary morphology failed (Laubichler 2003; Nyhart 1995; Nyhart 2002). Modern Evo Devo, on the other hand, does provide empirically verifiable hypotheses about the role of developmental mechanisms in evolutionary change, as well as a mechanistic explanation of homology. It is, therefore, in one sense a continuation of the nineteenth-century program of evolutionary morphology.

The story of how problems with the comparative and descriptive approaches contributed to the rise of experimental biology has been told several times (Allen 1975; Coleman 1971; Maienschein 1991). For our problem of understanding the relationship between form and function, two aspects of this shift are important. (1) With regard to development, the emphasis shifted to mechanistic models and explanations. This shift implied a reversal in the explanandum of embryology. While within the framework of comparative anatomy and embryology specific forms were explained by embryological evidence, now the patterns and processes of embryology were themselves the focus of scientific attention. This shift in emphasis brought with it more of a functional orientation into developmental biology, as the roles of various causal agents in development were investigated. (2) With regard to physiology, the experimental ideal in biology broadened this traditional approach to include many more species and a comparative dimension. All of this led to a higher visibility of functional considerations within biology that ultimately revealed the mechanistic basis of many adaptations in both animals and plants.

FORM AND FUNCTION AND THE MULTIPLE ORIGINS OF TWENTIETH-CENTURY EVO DEVO

The turn of the twentieth century was characterized by several developments of interest to our questions about the relationship between form and function. The rapid progress in experimental biology yielded several new perspectives on form (ontogenetic development) and function (physiology), while the rediscovery of Mendel's laws, and the subsequent emergence of the science of genetics, contributed more than any other scientific development to the separation of the traditional problem of generation into three distinct areas of investigation – inheritance, development, and evolution (Allen 1975; Laubichler and Maienschein 2007a). Furthermore, in part as an antithesis to what some perceived as overly reductionist tendencies within experimental biology, a renewed interest in form led to several new proposals, ranging from D'Arcy Thompson's version of mathematical biology to discussions within paleontology and morphology that are now considered part of the revival of "idealistic morphology" (Meyer 1926; Naef 1919; Thompson 1917; Troll 1928). Many of these proposals were attempts to fill the void left by the eclipse of Darwinism (Bowler 1983). Around 1900, many biologists had become dissatisfied with natural selection as the main mechanism of organic evolution. Instead, they argued that additional "internal" forces are needed to understand the specific patterns of phenotypic evolution. Some of these proposals focused on the problem of inheritance, suggesting various mechanisms of how the environment could influence the hereditary material; others emphasized the apparently directed patterns of phenotypic transformations (especially within the fossil record) and argued that the action of internal mechanisms are required to explain such orthogenetic series.

The scientific problems that many of these suggestions attempted to address were the origin and patterns of phenotypic variation. Variation is the raw material for natural selection; but what is the source of variation, and is variation random, constrained, or even directed? Darwin's neo-Lamarckian leanings are well known; and it is equally clear that these represent Darwin's attempt to deal with the problem of variation. Where Darwin was focusing on a direct influence of the environment, many others have been looking to development for an answer to this question. These suggestions, proposed in the course of the last century, thus represent several earlier attempts at what we now call Evo Devo.

Several proposals involve a broader conception of the gene than that which has dominated twentieth-century evolutionary biology. Among these, the school of developmental physiological genetics, which includes among others Richard Goldschmidt and Alfred Kühn, emphasized the developmental role of genes (Laubichler and Maienschein 2007b; Laubichler and Rheinberger 2004). Their difficult experimental approaches were designed to identify the function of genes in development. But they also proposed several mechanistic models of how developmental systems produce phenotypes, and how changes within these systems contribute to corresponding changes in phenotypes. Developmental physiological genetics of the first half of the twentieth century is thus a legitimate predecessor of Evo Devo.

Richard Goldschmidt, one of the proponents of this approach, is arguably the most important figure of the Modern Synthesis in the sense that he provided Dobzhansky and others with a common enemy, whose heretical ideas helped them shape their own conceptions of evolutionary change. As is well known, developmental biologists were generally not part of these deliberations (whether because of an active exclusion or a lack of interest is not yet clear) (Mayr and Provine 1980). But as a consequence, the Modern Synthesis view is that evolutionary processes are driven largely by (random) genetic changes on the one hand and by functional interactions of organisms with their environment on the other. Development, which would mediate between genes and phenotypes, is considered a black box that does not alter the fundamental dynamics of natural selection. In this conception, evolutionary change and the observed patterns of phenotypic variation are driven by function (fitness).

This view of the evolutionary synthesis, especially in its so-called "hardened" form, was soon challenged, first by individuals trained in fields such as morphology, paleontology, comparative embryology, and developmental biology. These individual voices of dissent would by the late 1970s and early 1980s be organized into a movement that would soon be known as evolutionary developmental biology, or Evo Devo (Hall 1992; Hall 1998).

Initially, these dissenters objected to individual claims and perceived shortcomings of the Modern Synthesis. Paleontologists argued that the implicit gradualism of evolutionary models does not correspond to observed patterns of the fossil record, and they suggested a whole range of macroevolutionary principles, including some, such as heterochrony, that had a developmental basis. Morphologists complained that the gene-centered perspective of the Modern Synthesis does not explain the structured hierarchy of forms and the nested nature of homologies. They also provided alternative theories of morphological evolution that included developmental principles, such as burden, or developmental constraints in explaining the conservation of specific characters. Developmental biologists also objected to the gene-centered view, arguing that the mechanisms of morphogenesis need to be an important part of any explanation of form. During this first period, what is now known as Evo Devo was represented by a variety of theoretical positions and concepts, which aimed to correct or expand the Modern Synthesis (Laubichler and Maienschein 2007b; Love and Raff 2003).

Ironically, it was the field of developmental genetics that provided Evo Devo with both a public relations boost and a possible new foundation that had the potential to bring together many of these important but largely diverse ideas. The discoveries of (1) the conservation of *Hox* genes and other developmentally relevant transcription factors; (2) the structured regulatory logic of developmental systems, including the complete description of the first set of regulatory gene networks; and (3) the limited number of structural genes in higher organisms have led to a re-evaluation of some of the basic Modern Synthesis assumptions about phenotypic evolution. The new paradigm emphasizes that in order to understand phenotypic evolution, one first has to understand the genetic toolkit of development and the nature of gene regulation in development. These two developmental concepts provide the basis for the study of variational properties of phenotypes, which are, after all, the raw material for natural selection to act upon.

With respect to understanding form and function, the scientific landscape of nineteenth- and twentieth-century biology was highly diverse; one can find arguments for almost any position, emphasizing the primacy of either form or function as the genuine explanatory principle of biology. This is not the place to reconstruct all these arguments, but this brief historical sketch should have made clear that the conceptual histories of the relationship between form and function and those of Evo Devo are closely intertwined. And, as I will argue below, the current incarnation of Evo Devo actually has a good chance of arriving at a conceptual solution to these long-standing problems.

FORM AND FUNCTION AND THE MAIN CONCEPTUAL QUESTIONS OF TWENTY-FIRST-CENTURY EVO DEVO

As we have seen, many early proponents of Evo Devo had initially presented their positions as alternatives to what they perceived as a "hardened" version of the Modern Synthesis in evolutionary biology – in a sense, presenting "form" as an alternative or antithesis to prevailing attitudes that privileged functional explanations. The current focus in Evo Devo, however, is mainly on how to accomplish integration across these different explanatory frameworks and intellectual traditions. But, as almost all participants in these discussions realize, synthesis talk is cheap; really accomplishing integration between different research traditions, especially those that have deep historical roots and involve complex experimental and theoretical traditions, is not easy (Hall 1998; Hall 2000; Laubichler 2007; Wagner 2000; Wagner 2001; Wagner and Larsson 2003). It is therefore not at all surprising that many Evo Devo researchers have emphasized the need for conceptual/philosophical and historical analysis as part of their ongoing efforts to realize the full potential of Evo Devo.

Summarizing these current discussions we can distinguish several different types of challenges in twenty-first-century Evo Devo: some have to do with the practical integration of separate research programs, including the selection of appropriate model organisms and research methodologies best suited to address genuine Evo Devo questions; while others are mainly connected to the problem of how the insights and results from different research traditions and methodologies – such as those that have historically emphasized "form" and those that have focused on "function" – can be integrated into a single coherent conceptual framework. Here, and in this volume more generally, we focus mostly on the issues connected with conceptual integration. However, the question "what is the right model organism for Evo Devo?" has recently received considerable attention, and it does influence in no small part the chances for accomplishing conceptual integration (Collins *et al.* 2007; Milinkovitch and Tzika 2007). We therefore also briefly discuss this issue.

Our brief historical overview clearly demonstrates that what we now call Evo Devo has many different roots, almost all connected in some fashion to the problem of form and function. These multiple origins of Evo Devo are also reflected in its several distinct research programs, illustrating the methodological pluralism of present-day Evo Devo, as well as its historically different questions and problems. This multitude of approaches and the wide range of topics also help us to understand the current prominence of Evo Devo-related research within many areas of biology. However, the very same diversity that makes Evo Devo research so productive also represents a formidable challenge to those who want to integrate these different perspectives within a new synthesis.

But before we discuss the possibilities of such an integration and its consequences for the problem of form and function, let us first briefly mention (1) the main research programs, (2) the dominant research questions, and (3) several of the fundamental concepts that characterize present-day Evo Devo (Hall 1998; Laubichler 2005; Laubichler 2007; Laubichler and Maienschein 2007b; Laubichler and Müller 2007b; Love and Raff 2003; Müller 2005; Müller 2007).

The main research programs in Evo Devo and their relation to form and function

(1) *The comparative program.* Comparative approaches are characteristic of most areas of biology insofar as these all deal with the diversity of organisms. Comparative approaches are also among the oldest within biology and have traditionally emphasized form on all levels of analysis – from morphology and anatomy to histology and embryology, and, today, also genomics and structural biology. These approaches continue to raise important questions and methodological challenges and also provide the foundation for all further investigations into biological processes or functions. In the context of Evo Devo, comparative perspectives apply to phenotypes (adult and embryological) as well as to gene expression patterns and their regulation.

A major part of the explanatory strategies of Evo Devo rests on the correlation between differences in gene expression patterns and phenotypic differences. Based on those structural differences of form, Evo Devo researchers attempt to develop causal models of phenotypic evolution. Thus, in a sense, they are integrating perspectives of form and function, as we will see in the discussion of one such model below. Furthermore, the comparative program is important for Evo Devo in that it is essential for establishing detailed and accurate phylogenies. Without such phylogenies, any number of genuine Evo Devo questions, including those about evolutionary transformations or homology, cannot be properly assessed.

(2) *The experimental program.* This part of Evo Devo continues many of the approaches of experimental embryology and developmental physiology that emerged during the first decades of the twentieth

century. The goal of all these approaches has been to elucidate the mechanisms of development. Over the last century this tradition has changed, mostly due to the introduction of new methods, which have allowed researchers to revisit age-old problems, such as morphogenesis, induction, and differentiation. All experimental approaches in Evo Devo also depend on the selection of the "right organisms for the job." And while traditional developmental biology has largely used the same seven canonical model organisms, Evo Devo researchers have in recent years adopted several new model organisms that are better suited for addressing the problems of phenotypic evolution (in some cases, this meant that they actually rediscovered model organisms that had been used during the early period of experimental embryology).

The reason that developmental biology continues to focus on its canonical experimental organisms is that these are well suited to genetic analysis, in line with the emphasis on the role of genes in development. However, there is now also an increasing awareness about the importance of epigenetic and environmental factors in morphogenesis and evolution, something that is better studied with less "constructed" model organisms and under more natural conditions (Müller and Newman 2005a; Müller and Newman 2005b). We have already seen that regulation has become one of the central concepts in describing both developmental systems and their evolution. In manipulating the internal parameters of developing systems, such as the number of cells in developing limb buds, as well as the environmental factors, such as temperature, the experimental program within Evo Devo has produced some interesting results. Among these are the re-creation of ancestral morphological patterns in limbs or the elucidation of the rules of digit reduction (Alberch and Gale 1983; Müller 1989). Other experimental approaches combine developmental with ecological and evolutionary concerns. Paul Brakefield's studies on the development, evolution, and ecological significance of butterfly eyespots are an outstanding example of such an integrative approach to Evo Devo.

All these experimental approaches contribute greatly to our understanding of the patterns and processes of evolutionary transformations. Many of them also address the problem of form and function. On the one hand, the development and evolution of phenotypic characters is the explanandum of experimental embryology; on the other hand, the experimental approach specifically addresses the function of various components of developmental systems.

(3) *The program of evolutionary developmental genetics*. To many in the larger community of biologists, evolutionary developmental genetics represents Evo Devo (Carroll *et al.* 2005). The discoveries of both the role and the conservation of the *Hox* genes and other transcription factors in the regulatory gene networks that control gene expression and the differentiation of embryonic anlagen have stimulated the field of developmental genetics. In addition, the high degrees of conservation of these genes, together with the recent (post human genome) findings about the small number of structural genes in vertebrate genomes, offered new perspectives. As a consequence of these findings, it has become clear that both embryonic differentiation and evolutionary transformation are largely a problem of regulation – that is, of differential gene expression during development and of differences in the regulatory architecture between species.

Among the ideas that have emerged in the context of these investigations are the notion of a *genetic toolkit for development* – a set of regulatory elements involved in the development of the main features of animal bodies, such as segmental patterning and axis formation. The idea brings with it the proposal to reconstruct a so-called *Urbilateria* as the ancestral condition of all higher animals, and the recognition that the structure of regulatory gene networks corresponds to the observed patterns of phylogeny. The latter combines the insights of the comparative program with the discovery of the genetic toolkit for development (Carroll *et al.* 2005; Davidson 2006a). The program in evolutionary developmental genetics has so far been largely typological, emphasizing questions of form more than those of function (other than the function of elements of the developmental system). But as more and more details of the regulatory gene networks are now known – revealing, for instance, the hierarchy of different elements of these networks – we begin to understand the phenotypic consequences of variation at all these levels. Based on these insights it is now feasible to combine a mechanistic understanding of development with traditional models of evolutionary dynamics, thus effectively integrating form and function in Evo Devo. On another note, the prominence of these approaches has contributed substantially to the false impression that Evo Devo began with the *Hox* story.

(4) *The theoretical and computational program within Evo Devo.* Not surprisingly, since its modern inception in the 1970s, Evo Devo has also triggered a lot of theoretical research and conceptual analysis. The theoretical program is now being consolidated, as newly developed formal, mathematical, and computational approaches begin to add more rigor to long-standing conceptual ideas. The theoretical program is also the domain to which most evolutionary biologists interested in Evo Devo contribute. There are, as yet, not many experimental Evo Devo approaches that study major phenotypic transformations, so a lot of the evolutionary focus of Evo Devo has been theoretical. This focus has been on new conceptual and mathematical ideas that include work on genotype-phenotype maps, including questions of how best to characterize phenotype and morphospace, theoretical and empirical analyses of modularity and robustness, phenotypic plasticity, life history, and evolvability. All these theoretical innovations contribute to a framework that might just prove flexible enough to integrate the different empirical and theoretical traditions of present-day Evo Devo (Callebaut and Rasskin-Gutman 2005; Hall and Olson 2003; Müller and Newman 2003; Schlosser and Wagner 2004; West-Eberhard 2003).

Another interesting aspect of the theoretical program is that Evo Devo researchers have interacted more closely with historians and philosophers of biology, especially with respect to the possible conceptual unification of otherwise diverse experimental approaches. The interdisciplinary orientation of the theoretical program also represents a counterweight to the program in evolutionary developmental genetics, which is largely dominated by researchers trained in molecular or developmental biology. The two research programs often follow different epistemological convictions and explanatory frameworks.

Simply put, developmental biologists tend to be more interested in structural and typological explanations based on molecular and cellular mechanisms, while evolutionary biologists focus more on dynamic processes on a population and species level (see also Hall 2000 for a discussion about the differences between Evo Devo and Devo Evo, and Amundson 2005 for a detailed account of the different epistemological and explanatory frameworks). It should therefore come as no surprise that the integration of these two approaches is anything but straightforward. Theoretical work is also greatly aided by new developments in computational methods and representations. The databases of the various genome projects have been indispensable for identifying developmentally active genes and establishing their evolutionary history. As functional annotations of genes in these databases increase and gene ontologies become more sophisticated, it will soon be possible to extract the kind of information about developmental genes that is necessary for a more detailed understanding of the evolution of developmental systems. What we have learned for *Hox* and related genes, how their expression domains shift in different species, and how this correlates with morphological changes, for instance, or what the consequences of certain duplication events have been, will soon be available for a large number of transcription factors, signaling genes, and receptor proteins. In addition, computational reconstructions of gene expression patterns in developing embryos will organize data in a way that will greatly aid mathematical modeling of developmental systems.

To sum up, the theoretical and computational program within Evo Devo is about to get a great boost from the increasing success of experimental approaches. Finally, within the theoretical program, the question of form and function is also addressed most clearly. For instance, Ron Amundson's analysis of the different epistemological frameworks of developmental and evolutionary biology clarifies the issue from a conceptual point of view and also suggests ways for a potential conceptual integration. We will sketch a mechanistic model that integrates the perspectives of form and function below, but it is already clear that this model has many conceptual and philosophical implications that need to be addressed (see also Hamilton, ch. 10 in this volume).

This brief sketch of some of the different research programs within Evo Devo and their relationship to the problem of form and function should have made it clear that this age-old problem of biology is still of central importance today, but also that Evo Devo's integrative perspective is more promising with regard to actually reaching a resolution than many earlier attempts. Another way to help us understand the relationship of Evo Devo to the problem of form and function is to investigate some of the concrete research problems that make up the core of present-day Evo Devo.

Research questions of Evo Devo

Presently, Evo Devo investigates a whole range of specific questions (Laubichler 2005; Laubichler 2007; Müller 2005; Müller 2007b; Wagner *et al.* 2000), several of which are specifically connected to the problem of form and function. Among those, the following six problem areas are especially relevant:

(1) *The origin and evolution of developmental systems.* Here the focus is on the evolutionary transformations of developmental systems. Developmental systems, like any other character of organisms, are subject to evolutionary forces. However, the transformations of these systems are especially relevant for explaining patterns of phenotypic evolution, as every phenotypic feature is the product of development. Research in this area has thus far revealed the modular architecture of developmental systems and regulatory gene networks, and currently investigates their robustness and how different developmental modules are combined and regulated. This set of questions touches directly on one aspect of the form-and-function relationship, as they focus on the mechanisms of generating different forms.

(2) *The problem of homology.* The problem of explaining homology has been called the central question of all biology. One prominent distinction is that between a historical homology concept, used mainly in phylogenetic reconstructions and investigations of the distribution of homologous characters, and a biological homology concept that attempts to provide an explanation of the very existence of homologies as a consequence of the developmental mechanisms that generate them. The Evo Devo framework employs both notions of homology and also tries to integrate them, explaining the specific distributions of homolog as a consequence of the conserved features of development. Again, this area of Evo Devo is mainly concerned with the problem of form.

(3) *The genotype–phenotype relation.* Mapping genotypes onto phenotypes has emerged as a main problem within evolutionary and quantitative genetics during recent decades. It is also the question that most directly combines the molecular details of developmental

mechanisms with formal models of evolutionary biology. For a long time, population genetic models assumed that development does not affect the mapping of genotypic onto phenotypic variation in any important way – development was thus treated as a constant. As this position can no longer be upheld, investigations into the formal properties of the genotype-phenotype map have become a major focus of the evolutionary branch of Evo Devo researchers. Within this area of Evo Devo, the analysis of form and function is integrated within a formal framework: in models of evolution, phenotypes are seen as the carriers of function (fitness), with the recognition that the developmental system – that is, the genotype-phenotype map – plays a major role in mediating between genotypic and phenotypic variation by introducing considerations of form (development) to the analysis of the evolutionary dynamics considerations of function (fitness).

(4) *The patterns of phenotypic variation.* This set of questions is closely connected to the genotype-phenotype map and the form and function problem. It has long been known that patterns of phenotypic variation are highly clustered and constrained. Explanations of this phenomenon have always included references to developmental systems, mostly in form of developmental constraints that limit the possible phenotypic variants. Recently, however, it has become clear that the ability of certain lineages to evolve (their evolvability) crucially depends on the existence of these developmental constraints, as these in fact facilitate variation for natural selection to act upon. Understanding these possibilities and limitations of phenotypic variation is thus crucial to the integration of form and function.

(5) *The role of the environment in development and evolution.* More recently, considerations of environmental factors have also been incorporated into the Evo Devo research program. The more we learn about the molecular mechanisms through which the environment can exert an influence on the expression of phenotypes, such as DNA methylation or endocrine disruptors, the more it becomes clear that the environment has to be part of explanations of the development and evolution of phenotypic variation. With respect to the form and function problem, this part of Evo Devo has the potential to overcome the internal (form) versus external (function) divide.

(6) *The origin of evolutionary novelties.* Explaining the origin of novel phenotypic traits has been one of the biggest challenges of evolutionary biology. It is also the main focus of Evo Devo, which offers an integrated perspective in addressing this problem. A mechanistic solution to the origin of evolutionary novelties problem is also one way in which Evo Devo approaches the form and function problem. We will discuss this question in more detail below.

This brief selection of research questions illustrates how Evo Devo's main focus is to explain the patterns and processes of phenotypic evolution. While at first glance this might seem a puzzling statement – has evolutionary biology not always been concerned with phenotypic evolution? so what is new and different about the Evo Devo approach? – Evo Devo does indeed represent a reorientation of evolutionary biology. The main difference is that phenotypes are no longer seen merely as the passive by-products of evolutionary processes that are mainly driven by internal genetic changes and external evolutionary forces, such as natural selection or genetic drift, but rather as the locus of integration of a whole range of mechanisms, from molecular and developmental to physiological and environmental. Consequently, the patterns and processes of phenotypic change result from a combination of all these diverse causal mechanisms from a variety of organizational (from genes to the environment) and temporal (developmental, life history, and evolution) scales. Evo Devo thus represents the latest attempt to integrate form and function within one explanatory framework.

For Evo Devo to be successful it has to overcome several obstacles, some of a deep conceptual nature, others more a consequence of historically entrenched habits of scientific practice. Conceptually, the different frameworks of organism-based experimental developmental biology and population-based evolutionary biology are arguably the largest stumbling block to an Evo Devo synthesis. This can be illustrated, for example, by the ways that both traditions approach efforts to understand the effects of a gene. In developmental genetics an effect of a gene is defined by its role in such processes as differentiation; the main emphasis, therefore, is on the regulation of gene expression. Furthermore, target genes in one system are often identified by what is known about their function in another organism – a strategy that has led to the discovery of high degrees of conservation of both sequence identity and function throughout the metazoa and higher plants. These discoveries also reinforced what has been called the typological approach of developmental genetics.

Evolutionary biology, on the other hand, emphasizes the role of genetic variation within and between populations. As long as both disciplines are still limited by their respective conceptual positions, it is difficult to accomplish an integrative theory of phenotypic evolution. For Evo Devo to be successful, it will have to develop conceptual innovations that overcome these incompatible views. One such conceptual innovation is the notion of regulatory evolution based on the known details of the molecular mechanisms of gene regulation and differentiation. This idea of regulatory evolution suggests a mechanistic understanding of phenotypic diversity from a developmental perspective. It also has the potential to be incorporated into models of evolutionary change, once the genetic variation governing these changes in regulatory networks has been identified. The notion of regulatory evolution thus shows that, based on conceptual analysis (the theoretical program of Evo Devo) and the availability of new data (the developmental genetics program of Evo Devo), it has been possible to detect the inconsistencies between different models of explanation that eventually forced a change in the conceptual framework of explanations of phenotypic evolution, including the problem of form and function.

Regulatory evolution is just one of a set of concepts and unifying themes that together make up the proposed integration of Evo Devo. Others include such notions as constraint, heterochrony, modularity, hierarchy, homology, evolvability, emergence, plasticity, innovation, robustness, and regulatory networks. What is of particular interest here is that the majority of these concepts or their predecessors originated in the context of earlier discussions about form and function, something we need to keep in mind when we explore Evo Devo's potential for achieving a successful integration of form and function.

So far we have discussed the pluralism of research programs and problems, but today's Evo Devo is also characterized by several unifying concepts that contribute significantly to a more integrated approach to the problem of form and function (see also Müller 2007a).

Unifying themes in Evo Devo and their connections to form and function

Among the many concepts that either originated within Evo Devo or have been transformed by Evo Devo perspectives, four are especially relevant for our discussion of form and function. These are regulation (including regulatory evolution and regulatory gene networks), modularity, plasticity (including a discussion of behavior as the ultimate integrator of form and function), and innovation. All reflect principles of organismal organizations, their development, and evolution. There are, of course, additional concepts that are part of the theoretical framework of Evo Devo. Some, like heterochrony, developmental constraints, burden, or hierarchy, have been part of the first wave of Evo Devo proposals. Others, such as robustness, evolvability, or homology, are the subjects of intense discussions today; but insofar as these are relevant to the resolution of the form and function problem, we will discuss them in the context of our four examples.

Regulation has long been recognized as a fundamental property of organisms as well as other biological and social systems. Indeed, the concept of regulation has been tied to the origin of both theoretical biology and philosophy of biology, dating all the way back to the late eighteenth century. Regulation is a function of organisms, and enables them to maintain their structural cohesiveness (form) in light of a variable environment. What is of special interest is that development, or the origin of organismal form, is also a regulative process. How else would it be possible that, despite unpredictable and variable environmental conditions and perturbations, the outcome of ontogeny is more or less predictable, that a chicken egg gives rise to a chicken and a sea urchin larva to a sea urchin? The regulatory capacity of developing embryos was also one of the first challenges to a simple deterministic or preformist theory of development that tried to explain differentiation as a consequence of an uneven distribution of hereditary material. Hans Driesch's famous experiment that demonstrated the regulatory potential of blastomeres thus gave rise to the dynamic conception of development that emphasizes the interaction between different parts of the system.

In today's Evo Devo, regulation is central to our understanding of both the development and the evolution of organisms. It is an essential part of our mechanistic theory of phenotypic evolution that provides an integrated perspective on the problem of form and function. The actions of gene regulatory networks provide us with a clear explanation of the problem of differentiation. Different cell types are ultimately characterized by the expression of different sets of genes, which are, in turn, a consequence of differential gene regulation. Uncovering the complex molecular machinery that regulates these expression patterns has been one of the major accomplishments of developmental genetics during recent decades.

In addition, it has become clear that regulatory networks not only act within the genome, but that different levels of phenotypic organization also have their own regulatory control mechanisms that guarantee their stability in light of variation of their parts. As we have found out more about the molecular details of development, it has become clear that (1) a large degree of conservation exists among the different elements of the developmental system; (2) there are many cases in which we find phenotypic stability despite substantial changes among the elements of the developmental systems – in other words, the same phenotype can be realized through the action of different molecular factors; and (3) these regulatory networks have their own conserved architectural structures that show different degrees of conservation. There are some elements that are extremely conserved and whose action is essential for any normal development ("kernels"), while others are more variable and can also be deployed in a variety of different contexts ("plug-ins" and "switches").

This last property of regulatory (gene) networks also helps us understand the developmental mechanisms of phenotypic evolution, which are now considered to be largely a consequence of regulatory evolution. One immediate consequence of the various genomics projects has been the recognition that there are fewer genetic differences between different species (such as *Drosophila* and human) than originally thought. These results imply that phenotypic differences and phenotypic evolution are more the result of changes in the expression patterns of genes than they are of novel genes. These findings undermined any simple causal model of phenotypic evolution that connects novel phenotypic characters with new genes. But they contribute substantially to the integration of form and function in the sense that it is now clear that phenotypic differences (form) are a consequence of differences in the function (regulation) of the developmental systems (form) that are selected because of their contributions to organismal fitness (function). The architecture of these regulatory (gene) networks is also an instance of modularity, another general principle of biological organization that is a prominent unifying theme within Evo Devo.

Modularity is a fundamental characteristic of all biological organizations; it is thus a universal principle of biological form. We find modules on all levels of organization, from the genome to cellular, anatomical, and behavioral structures. In addition to structural modules, there are also dynamic modules such as specific developmental processes that remain stable across a wide range of different organisms, such as certain elements of regulatory networks, signaling cascades or processes of induction. Modularity is also closely connected to the concept of homology and the basic principles of phenotypic organization.

Several major research questions of Evo Devo relevant to the form-and-function problem are directly connected to issues of modularity. Most models of adaptation or optimization of individual organismal character depend on what Herbert Simon called near-decomposability – in other words, on the existence of both phenotypic and corresponding genotypic modules. Genotypic modules are characterized by reduced levels of pleiotropy – that is, the genes connected to the particular phenotype under selection should not also affect too many additional phenotypic features, as this would greatly limit the ability of natural selection to optimize any given function. Expressed in terms of the genotype-phenotype map, this implies that, due to the modular architecture of this map, it is possible to decompose it into several smaller, independent maps of lower dimension that can each then be optimized independently by natural selection. Since such a structure is itself an adaptive feature of organismal design – it does, after all, allow for selective optimization of traits – it is reasonable to assume that the modular structure of the genotype-phenotype map has itself been the product of evolutionary change. We have already mentioned that the genotype-phenotype map is a formal representation of the developmental system. Therefore, the observed modularity of organismal design is a consequence of regulatory evolution of the developmental system, which is, as we have seen above, directly linked to the mechanistic integration of form and function in the context of Evo Devo.

Phenotypic plasticity is the capacity of a single genotype to generate different phenotypes in response to variable environmental conditions. Plasticity is thus a major functional response of organisms to cope with their unpredictable environment. As such, it is another case for the regulatory potential of organisms. Research into the mechanisms of phenotypic plasticity has recently emerged as a major theme within Evo Devo, especially in light of recent efforts to add an ecological dimension to its portfolio. Phenotypic plasticity is also particularly interesting for our discussion as it connects form and function in multiple ways.

One organismal property that displays high degrees of phenotypic plasticity is behavior. Studying behavior, and especially social behavior, from an Evo Devo perspective has become an active area of research during the last few years. Social systems are interesting as they are built from linear and non-linear behavioral interactions among individual organisms involving communication (signaling) and stimulus–response systems. They develop at the level of the individual through ontogenetic processes, and at the social-system, or superorganismal, level through growth and differentiation of the social group. Social systems, such as those represented by social insects, are also the product of evolution, and indeed, the emergence of social behavior is considered one of the key innovations in the history of life. We also know that there are allelic substitutions at variable gene loci that affect developmental events, with downstream effects on interactions and response systems at the level of individuals that result in adaptive social patterns and emergent colony-level traits. However, in order to differentially affect behavior, allelic differences must be translated into differential expression or function of peptides and proteins that themselves are parts of signaling systems, and which in turn form control systems that cause changes in the signaling and stimulus–response systems of individuals. In other words, the function of these allelic variants is determined by both the developmental and the social systems.

Furthermore, these control systems directly affect and are affected by developmental processes in both immature and adult insects, and by the developing colony social structure. Behavior is thus the product of complex regulatory processes during ontogeny, phylogeny, and the life history of individuals, and, in the case of social organisms, the colony or social system as well. The advantages of studying social behavior from an Evo Devo perspective are manifold and are of a pragmatic (see below for a discussion of social insects as a model system for Evo Devo) as well as a theoretical nature. Conceptually, behavior, as the ultimate integrator of form and function, helps us to expand the framework of regulatory gene networks to include organismal (physiology) as well as colony-level and environmental effects.

Innovation, or the origin of evolutionary novelties has often been identified as the genuine Evo Devo problem. We have already seen that recent evidence strongly suggests that any new phenotypic feature is the consequence of one form or another of regulatory evolution – that is that the developmental system determines whether or not a new phenotype is produced in the first place. Natural selection, of course, then decides its future fate. Innovation is thus connected to what Darwin already identified as the problem of the origin of variation. But while previous suggestions often invoked some sort of special cause or mechanism, such as macro- or systemic mutations, in order to account for phenotypic innovations, within the Evo Devo framework these features are seen as a consequence of regulatory or developmental evolution. Understanding the limits and possibilities of the developmental systems thus becomes a major part in any explanation of macroevolution. The limits and possibilities of evolutionary innovations are also closely tied to understanding the relationship between form and function: what forms are possible determines which functions an organism can perform; and how well it performs a function, in turn, decides which variants will survive. But both are in a sense enabled by the underlying mechanisms of the developmental system. Thus within an Evo Devo approach, the chicken-and-egg problem of the primacy of form or function dissolves.

How selecting the right model systems matters for understanding form and function

One recurring concern of Evo Devo is how to select the best-suited model system for investigating the role of development in phenotypic evolution (Collins *et al.* 2007; Milinkovitch and Tzika 2007). Developmental biology is still largely based on the canonical seven model organisms, and these are in many ways not optimal for addressing Evo Devo problems. What, then, are the properties of a good model system in Evo Devo? And, specifically, how are these connected to the problem of integrating perspectives of form and function?

For example, the papers in this volume represent a wide variety of taxa – from land plants to butterflies, sea urchins, fish, and birds – and combine multiple investigative methods – formal mathematical analysis, comparative studies of gene expression, functional analysis, and selection experiments – in order to address a variety of questions connected with the origin of novel phenotypes. In selecting the appropriate organisms for each of these approaches, two sets of demands have to be met: (1) pragmatic considerations of experimental manipulability, coupled with (2) theoretical considerations related to whether or not a specific model organism is representative for the phenomenon in question. These two demands sometimes conflict. Easy manipulability facilitates experimental work and the standardization of results, thus ensuring the quality and comparability of the experimental data. Yet that raises questions about the degree to which any particular organism is representative. Another issue is whether it is actually possible to develop a more general model or theory based on work done with one or a few specific model organisms. Unfortunately, this problem may well be more acute in the context of Evo Devo than it is in many other areas of biology (Collins *et al.* 2007; Metscher and Ahlberg 1999).

Experimental developmental biology is to a large degree still centered around seven basic model systems: the fruit fly *Drosophila melanogaster*; the nematode *Caenorhabditis elegans*; the mouse *Mus musculis*, the frog *Xenopus laevis*; the zebrafish *Danio rerio*; the chick *Gallus gallus*; and the mustard *Arabidopsis thaliana*. However, a number of other organisms are beginning to emerge, or in some cases re-emerge, such as sea urchins or ascidians. Traditionally, these organisms were selected as model systems because of their easy manipulability and, except for the chick, because they turned out to be well suited for research within the genetic paradigm in developmental biology.

Unfortunately, these organisms are not particularly useful for Evo Devo questions. Most are highly derived and specialized, and thus not suited to addressing major phenotypic transformations or any of the other questions that are being asked in Evo Devo with respect to form and function. It is by now widely accepted that addressing these problems requires new model systems. The emphasis on genetic approaches, for instance, contributed to the diminished role of sea urchins in mid-twentieth-century developmental biology. Today, sea urchins are again being used, largely because new methods of molecular biology have allowed the purple sea urchin, *Strongylocentrotus purpuratus*, to become the paradigmatic model for the study of gene regulatory networks (Davidson 2006b; Davidson *et al.* 2002). In addition, one of the earliest and by now best-known Evo Devo model systems involves two different sea urchin species of the genus *Erythrogramma*. This system was developed by Rudy Raff and co-workers to study the differences between larval and direct development in closely related species (Raff and Wray 1989). Now, as Evo Devo has matured and developed its own set of research questions related to form and function and the origin of novel phenotypes, new model systems are rapidly emerging.

Evo Devo investigators are currently introducing several new model systems, as well as "reappropriating" some of the model systems of developmental biology, such as the zebrafish and *Drosophila*. These two allow for the fine-grained comparative studies needed for the discovery of specific differences within and between lineages by studying altered gene expression patterns on a microevolutionary scale. The zebrafish with its sequenced genome, for instance, has become a useful source of candidate genes through which the evolution of specific piscine lineages might be studied (Webb and Schilling 2006), while *Drosophila* has allowed David Stern to detect the developmental basis for within-population variation in multiple species of *Drosophila* (Stern 2003; Stern 2006; Stern 2007).

The dog *Canis familiaris* (Neff and Rine 2006) and the three-spined stickleback are also considered well-suited model systems for studying the consequences of altered gene expression during evolution. These are new model systems, specifically developed in the context of Evo Devo to identify genes whose small changes can make large phenotypic differences. Other new and non-traditional organisms are the cnidarian *Nematostella*, as a model system for looking at the origins of the bilateria (Martindale *et al.* 2004), and the dung beetle *Onthophagus*, which is proposed as a model system for studying the evolution and properties of developmental plasticity (Emlen 2000). In this regard, it is important to note that "model systems" in Evo Devo are not merely "model organisms." Rather, these systems include the organism plus the historical or ecological context of the organism (Collins *et al.* 2007; Gilbert 2001).

As should be clear by now, the "hybrid" character of Evo Devo, with its emphasis on the integration of developmental and evolutionary perspectives, actually provides a formidable challenge for the selection of model organisms. The perfect system would be one for which we already know a good deal about the molecular mechanisms of development, ideally including a completed genome sequence; that is amenable to molecular and genetic analysis; that has a range of well-defined phenotypic innovations for which a good phylogeny exists; and that has closely related sister taxa that do not possess the phenotypic innovations in question, and for which we know their ecology well enough to be able to formulate testable hypotheses about the selective advantage of these phenotypic innovations.

One of the emerging new model systems in Evo Devo that fulfills all these requirements is the social insects, especially the honey bee, *Apis mellifera* and, as a whole group, the ants (Amdam *et al.* 2006; Laubichler and Gadau in press; Page 1997; Page and Amdam 2007; Toth and Robinson 2007). An Evo Devo perspective on Social Insects is especially interesting since the evolution of sociality – from solitary to eusocial – is often seen as one of the major transitions in evolution. Such transitions involve the emergence of a whole new set of novel behaviors and associated phenotypic innovations, thereby dramatically changing the evolutionary dynamics of the group. An organized Evo Devo approach to social insects is only now emerging, but it has already expanded traditional Evo Devo perspectives through its emphasis on the development and evolution of behavior. Behavior is an especially interesting phenotype as it acts as the ultimate integrator of form and function. Biological functions are always dynamic, and behavior mediates between an individual and its environment as well as between different individuals. But behavior (function) is, of course, enabled and constrained by the morphological properties of organisms (form). In addition, specific behaviors are stable characters or homologies that can be recognized as the same between different species. Furthermore, behavior is generally more variable than other features of phenotypic plasticity. Therefore, social insects offer a whole range of new perspectives for Evo Devo. And they can also serve as a viable test case for the proposed mechanistic solution to the problem of form and function.

CONCLUSION: A PROPOSED MECHANISTIC SOLUTION TO THE PROBLEM OF INTEGRATING FORM AND FUNCTION IN EVO DEVO

As we have seen, twenty-first-century Evo Devo offers a promising conceptual framework for addressing the relationship between form and function. It also ties the current issues, represented by the various research programs and concepts discussed above, to the rich and significant history. In a world that is largely composed of shades of gray, conceptual dichotomies, such as form and function, ontogeny and phylogeny, constraint and selection, micro- and macroevolution, provide convenient guideposts for structuring our understanding of these problems, as long as they do not harden into ontological alternatives, but rather remain fluid and dynamic. With regard to Evo Devo's challenge of integrating developmental and evolutionary processes and approaches, the form and function dichotomy offers an analytical perspective that helps to illuminate the challenges and possibilities of such integration.

We have already identified the main challenge of Evo Devo as connecting different temporal scales (ontogeny and phylogeny), perspectives (molecular interactions during development and variational properties of populations), and analytical traditions (experimental and mathematical). Using the form-versus-function dichotomy as an abstract schema, together with the case studies collected in this volume, allows us to sketch a conceptual model for the successful integration of evolution and development. An idealized model of evolution by means of natural selection (one that arguably nobody subscribes to in its extreme form) emphasizes the functional properties of organisms within populations. Insofar as phenotypic variation translates into differences in fitness, which in turn are due to differences in function, natural selection then leads to adaptation. Evolutionary change, in this view, would thus be driven by function. Any additional factor, such as constraints on the variational properties of phenotypes, would then be merely a modifying factor of an otherwise functionally driven dynamics.

An equally cartoonish version of developmental biology helps us highlight the differences between developmental and evolutionary approaches. The ultimate goal of developmental biology is to understand the patterns and processes connected to growth, differentiation, and morphogenesis. Therefore, developmental biology is largely an experimental science based on the analysis of standardized model organisms. Its goal, however, is to identify general principles that hold true for a large number of different species. This focus on generality leads necessarily to a typological orientation, a fact that has only been reinforced by the recent discoveries of the enormous amount of conservation of key elements of developmental systems, such as the *Hox* genes, other transcription factors, and the structure of gene regulatory networks (Davidson 2006a; Davidson 2006b; Davidson *et al.* 2003; Hinman *et al.* 2003; Howard and Davidson 2004; Levine and Davidson 2005; Manzanares *et al.* 2000; Negre *et al.* 2005; Santini *et al.* 2003).

These discoveries have led to the idea of a general toolkit for development, shared by organisms from rather distinct phyla (Carroll *et al.* 2001; Carroll *et al.* 2005; Davidson 2006a). The emphasis within these kinds of developmental explanations has thus been on form – or rather the (conserved) structure – of the developmental system. As a rule of logic, phenotypic innovations or variations would have to be explained by corresponding changes in the developmental systems. But until recently, the model-organism-based focus of developmental biology did not yield the kind of data – namely variational data within populations, or at the very least between closely related species – that would have been needed to integrate developmental with evolutionary processes in explanations of phenotypic evolution (Davidson 2006b; Kruglyak and Stern 2007; McGregor *et al.* 2007; Orgogozo *et al.* 2006; Stern 2003; Stern 2007; Sucena *et al.* 2003).

The genomic and post-genomic revolutions have added yet another twist to these attempts at integrating development with evolution. It has become clear that novel phenotypes are, for the most part, not primarily a consequence of novel genes, but rather of differences in the regulation of gene expression (in space, time, and quantity). Furthermore, several long-running research programs have revealed many details of the regulatory logic underlying gene expression and differentiation (Davidson and Levine 2003; Harafuji *et al.* 2002; Levine 1999; Levine and Davidson 2005; Levine and Tjian 2003; Mannervik *et al.* 1999; Markstein and Levine 2002; Markstein *et al.* 2002; Shi *et al.* 2005; Stathopoulos and Levine 2002; Stathopoulos and Levine 2004).

The emerging picture is one of structured complexity. The regulatory networks are often quite complex, again undermining any simple correlation between genes and phenotypes. But they nevertheless have a modular and hierarchical structure that allows us to correlate different aspects of network architecture with different types of variational properties among organisms. Based on these results, Eric Davidson and Douglas Erwin have suggested that different elements of the network architecture are connected with different morphological features of body plans that show varying degrees of conservation (Davidson 2006a; Davidson and Erwin 2006). Their model, initially derived from the molecular analysis of the genomic control of differentiation, thus provides the mechanistic underpinning for older observations in comparative anatomy and morphology that also described the nested hierarchy of body-plan features and homologies (Davidson and Erwin 2006; Riedl 1975; Wimsatt 2007). Modern developmental genetics has thus led us to a molecular understanding of the fundamental principles of form.

It also suggests a road map for accomplishing a true integration of developmental and evolutionary processes, as well as the integration of the age-old perspectives of form and function. What the molecular analysis of developmental processes and regulatory gene networks provides is a mechanistic understanding of both the development and evolution of phenotypic characters. On the developmental side, the analysis of these mechanisms has revealed general architectural principles of developmental systems, including the properties of regulatory gene networks, the logic of differentiation, and the properties of morphogenetic processes. Comparing these mechanisms across a range of different taxa has provided us not only with well-established correlations between the variations in gene expression patterns and phenotypic characters, but also with several concrete hypotheses about how changes in the developmental systems can explain evolutionary transformations of phenotypic characters, and about the evolutionary mechanisms driving these phenotypic changes (Arthur 2001; Baratte *et al.* 2006; Beldade *et al.* 2005; Beldade *et al.* 2007; Brakefield 2007; Cameron *et al.* 1998; Cebra-Thomas *et al.* 2005; Davidson and Levine 2003; Hansen 2006; Nijhout 2003; Peterson *et al.* 2000; Prum 2005; Salazar-Ciudad 2006).

The mechanistic framework of evolutionary developmental biology thus allows us to (1) understand the general principles of developmental systems governing phenotypic patterns; (2) identify the causal connections between variations in those developmental systems and observed patterns of phenotypic variation; (3) suggest concrete evolutionary scenarios of how underlying developmental changes govern evolutionary transformations of phenotypes; and (4) develop experimental tests to uncover the selective forces driving these evolutionary transformations. In this way, the mechanistic framework of developmental evolution unites the perspectives of form and function that have so often been associated with separate explanatory frameworks and sub-disciplines.

The main conclusions from these discussions are that (1) questions about structure, organization, and history – those themes that are traditionally associated with the concept of form – are a major research emphasis within current Evo Devo; (2) questions about function and purpose, both proximate and ultimate, still provide an important explanatory framework in the context of evolutionary explanations; (3) revolutionary implications of Evo Devo lie in the conceptual synthesis of considerations of both form and function through the mechanistic integration of developmental and evolutionary processes; (4) this synthesis is built around a set of concepts, such as modularity, hierarchy, constraints, evolvability, novelty and emergence, robustness, history, and so on; and (5) reaching the promise of Evo Devo will require not just continuous experimental work and more sophisticated mathematical models, but also a thorough analysis of the conceptual structure and a critical reading of the historical developments of the field. All the papers in this volume contribute to such interdisciplinary efforts, which have been essential for reaching the current level of success and excitement in Evo Devo. They will be even more important for reaching its full potential.

REFERENCES

Alberch, P. and Gale, E.A. (1983). Size dependence during the development of the amphibian foot. Colchicine-induced digital loss and reduction. *J Embryol Exp Morphol* 76, 177–97.

Allen, G.E. (1975). *Life Science in the Twentieth Century*. New York: Wiley.

Amdam, G.V., Csondes, A., Fondrk, M.K., and Page, R.E., Jr. (2006). Complex social behaviour derived from maternal reproductive traits. *Nature* 439, 76–8.
Amundson, R. (2005). *The Changing Role of the Embryo in Evolutionary Thought: Roots of Evo-Devo*. Cambridge and New York: Cambridge University Press.
Appel, T.A. (1987). *The Cuvier-Geoffroy Debate: French Biology in the Decades before Darwin*. New York: Oxford University Press.
Arthur, W. (2001). Developmental drive: an important determinant of the direction of phenotypic evolution. *Evol Dev* 3, 271–8.
(2002). The emerging conceptual framework of evolutionary developmental biology. *Nature* 415, 757–64.
Ballauff, T. and Ungerer, E. (1954). *Die Wissenschaft vom Leben; eine Geschichte der Biologie*. Freiburg: K. Alber.
Baratte, S., Peeters, C., and Deutsch, J.S. (2006). Testing homology with morphology, development and gene expression: sex-specific thoracic appendages of the ant *Diacamma*. *Evol Dev* 8, 433–45.
Beldade, P., Brakefield, P.M., and Long, A.D. (2005). Generating phenotypic variation: prospects from "evo-devo" research on *Bicyclus anynana* wing patterns. *Evol Dev* 7, 101–7.
Beldade, P., French, V., and Brakefield, P.M. (2007). Developmental and genetic mechanisms for evolutionary diversification of serial repeats: eyespot size in *Bicyclus anynana* butterflies. *J Exp Zoolog B Mol Dev Evol* 310B, 191–201.
Bonner, J.T. (1958). *The Evolution of Development: Three Special Lectures Given at University College, London*. Cambridge University Press.
Bowler, P.J. (1983). *The Eclipse of Darwinism : Anti-Darwinian Evolution Theories in the Decades around 1900*. Baltimore: Johns Hopkins University Press.
(1996). *Life's Splendid Drama: Evolutionary Biology and the Reconstruction of Life's Ancestry, 1860–1940*. University of Chicago Press.
(2003). *Evolution: The History of an Idea*. Berkeley: University of California Press.
Bowler, P.J. and Morus, I.R. (2005). *Making Modern Science: A Historical Survey*. University of Chicago Press.
Brakefield, P.M. (2007). Butterfly eyespot patterns and how evolutionary tinkering yields diversity. *Novartis Found Symp* 284, 90–101; discussion 101–15.
Breidbach, O. (2006). *Goethes Metamorphosenlehre*. Munich: W. Fink.
Breidbach, O., Fliedner, H.-J., and Ries, K. (2001). *Lorenz Oken (1779–1851): ein politischer Naturphilosoph*. Weimar: Verlag Hermann Böhlaus Nachf.
Breidbach, O., and Ziche, P. (2001). *Naturwissenschaften um 1800: Wissenschaftskultur in Jena-Weimar*. Weimar: H. Böhlaus Nachfolger.
Callebaut, W. and Rasskin-Gutman, D. (2005). *Modularity: Understanding the Development and Evolution of Natural Complex Systems*. Cambridge, MA: MIT Press.
Cameron, R.A., Peterson, K.J., and Davidson, E.H. (1998). Developmental gene regulation and the evolution of large animal body plans. *American Zoologist* 38, 609–20.

Carroll, S.B., Grenier, J.K., and Weatherbee, S.D. (2001). *From DNA to Diversity: Molecular Genetics and the Evolution of Animal Design*. Malden, MA: Blackwell Science.

(2005). *From DNA to Diversity: Molecular Genetics and the Evolution of Animal Design*. Malden, MA: Blackwell Publishing, 2nd edn.

Cassirer, E. (1950). *The Problem of Knowledge: Philosophy, Science, and History since Hegel*. New Haven: Yale University Press.

Cebra-Thomas, J., Tan, F., Sistla, S., Estes, E., Bender, G., Kim, C., Riccio, P., and Gilbert, S.F. (2005). How the turtle forms its shell: a paracrine hypothesis of carapace formation. *J Exp Zoolog B Mol Dev Evol* 304, 558–69.

Coleman, W. (1971). *Biology in the Nineteenth Century: Problems of Form, Function, and Transformation*. New York: Wiley.

Collins, J.P., Gilbert, S.F., Laubichler, M.D., and Müller, G.B. (2007). Modeling in Evo Devo: how to integrate development, evolution, and ecology. In M.D. Laubichler and G.B. Müller (eds.), *Modeling Biology: Structures, Evolution, behavior*. Cambridge, MA: MIT Press, pp. 355–78.

Davidson, B. and Levine, M. (2003). Evolutionary origins of the vertebrate heart: specification of the cardiac lineage in *Ciona intestinalis*. *Proc Natl Acad Sci USA* 100, 11469–73.

Davidson, E.H. (2006a). *The Regulatory Genome: Gene Regulatory Networks in Development and Evolution*. Burlington, MA: Academic Press.

(2006b). The sea urchin genome: where will it lead us? *Science* 314, 939–40.

Davidson, E.H., and Erwin, D.H. (2006). Gene regulatory networks and the evolution of animal body plans. *Science* 311, 796–800.

Davidson, E.H., McCay, D.R., and Hood, L. (2003). Regulatory gene networks and the properties of the developmental process. *Proc Nat Acad Sci of USA* 100, 1475–80.

Davidson, E.H., Rast, J.P., Oliveri, P., Ransick, A., Calestani, C., Yuh, C.H., Minokawa, T., Amore, G., Hinman, V., Arenas-Mena, C., Otim, O., Brown, C.T., Livi, C.B., Lee, P.Y., Revilla, R., Rust, A.G., Pan, Z.J., Schilstra, M. J., Clarke, P.J.C., Arnone, M.I., Rowen, L., Cameron, R.A., McClay, D.R., Hood, L., and Bolouri, H. (2002). A genomic regulatory network for development. *Science* 295, 1669–78.

Gilbert, S.F. (2001). Ecological developmental biology: developmental biology meets the real world. *Dev Bio* 233, 1–12.

Goethe, J.W.von (1824). *Zur Naturwissenschaft überhaupt, besonders zur Morphologie: Erfahrung, Betrachtung, Folgerung, durch Lebensereignisse verbunden*. Stuttgart: J.G. Cotta.

Goethe, J.W.von and Günther, H. (1981). *Anschauendes Denken: Goethes Schriften zur Naturwissenschaft*. Frankfurt am Main: Insel.

Gould, S.J. (1977). *Ontogeny and Phylogeny*. Cambridge, MA: Belknap Press of Harvard University Press.

Hall, B.K. (1992). *Evolutionary Developmental Biology*. London and New York: Chapman and Hall.

(1998). *Evolutionary Developmental Biology*. London and New York: Chapman and Hall.

(2000). Evo-devo or devo-evo – does it matter? *Evol Dev* 2, 177–8.

Hall, B.K. and Olson, W.M. (2003). *Keywords and Concepts in Evolutionary Developmental Biology*. Cambridge, MA: Harvard University Press.

Hansen, T.F. (2006). The evolution of genetic architecture. *Annual Review of Ecology, Evolution, and Systematics* 37, 123–57.

Harafuji, N., Keys, D.N., and Levine, M. (2002). Genome-wide identification of tissue-specific enhancers in the *Ciona* tadpole. *Proc Natl Acad Sci USA* 99, 6802–5.

Hinman, V.F., Nguyen, A.T., Cameron, R.A., and Davidson, E.H. (2003). Developmental gene regulatory network architecture across 500 million years of echinoderm evolution. *Proc Natl Acad Sci USA* 100, 13356–61.

Howard, M.L. and Davidson, E.H. (2004). Cis-regulatory control circuits in development. *Dev Biol* 271, 109–18.

Jahn, I. (1998). *Geschichte der Biologie: Theorien, Methoden, Institutionen, Kurzbiographien*. Jena: G. Fischer.

Kohn, D. and Kottler, M.J. (1985). *The Darwinian Heritage: Including Proceedings of the Charles Darwin Centenary Conference, Florence Center for the History and Philosophy of Science, June 1982*. Princeton, NJ: Princeton University Press, in association with Nova Pacifica.

Kruglyak, L. and Stern, D.L. (2007). Evolution. An embarrassment of switches. *Science* 317, 758–9.

Laubichler, M.D. (2003). Carl Gegenbaur (1826–1903): integrating comparative anatomy and embryology. *J Exp Zoolog B Mol Dev Evol* 300, 23–31.

(2005). Evolutionäre Entwicklungsbiologie. In U. Krohs and G. Toepfer (eds.), *Einführung in die Philosophie der Biologie*. Frankfurt am Main: Suhrkamp, pp. 322–37.

(2007). Evolutionary developmental biology. In D. Hull and M. Ruse (eds.), *Cambridge Companion to the Philosophy of Biology*. Cambridge University Press, pp. 342–60.

Laubichler, M.D. and Gadau, J. (in press). Social insects as models for Evo Devo. In J. Gadau and J. Fewell (eds.), *Organization of Insect Societies – From Genomes to Socio-Complexity*. Cambridge, MA: Harvard University Press.

Laubichler, M.D. and Maienschein, J. (2004). Development. In M.C. Horowitz (ed.), *The New Dictionary of the History of Ideas*. New York: Charles Scribner's Sons, vol. II, pp. 570–4.

(2007a). Embryos, cells, genes, and organisms: a few reflections on the history of evolutionary developmental biology. In R. Brandon and R. Sansom (eds.), *Integrating Evolution and Development: From Theory to Practice*. Cambridge, MA: MIT Press, pp. 1–24.

(2007b). *From Embryology to Evo-Devo: A History of Developmental Evolution*. Cambridge, MA: MIT Press.

Laubichler, M. and Müller, G.B. (2007). *Modeling Biology: Structures, Behavior, Evolution*. Cambridge, MA: MIT Press.

Laubichler, M.D. and Rheinberger, H.J. (2004). Alfred Kuhn (1885–1968) and developmental evolution. *J Exp Zoolog B Mol Dev Evol* 302B 103–10.

Levine, M. (1999). Transcriptional control of *Drosophila embryogenesis*. *Harvey Lect* 95, 67–83.

Levine, M. and Davidson, E.H. (2005). Gene regulatory networks for development. *Proc Natl Acad Sci USA* 102, 4936–42.

Levine, M. and Tjian, R. (2003). Transcription regulation and animal diversity. *Nature* 424, 147–51.

Love, A.C. and Raff, R.A. (2003). Knowing your ancestors: themes in the history of evo-devo. *Evol Dev* 5, 327–30.

Maienschein, J. (1991). *Transforming Traditions in American Biology, 1880–1915*. Baltimore: Johns Hopkins University Press.

Mannervik, M., Nibu, Y., Zhang, H., and Levine, M. (1999). Transcriptional coregulators in development. *Science* 284, 606–9.

Manzanares, M., Wada, H., Itasaki, N., Trainor, P.A., Krumlauf, R., and Holland, P.W. (2000). Conservation and elaboration of *Hox* gene regulation during evolution of the vertebrate head. *Nature* 408, 854–7.

Markstein, M. and Levine, M. (2002). Decoding cis-regulatory DNAs in the *Drosophila* genome. *Curr Opin Genet Dev* 12, 601–6.

Markstein, M., Markstein, P., Markstein, V., and Levine, M.S. (2002). Genome-wide analysis of clustered Dorsal binding sites identifies putative target genes in the *Drosophila* embryo. *Proc Natl Acad Sci USA* 99, 763–8.

Mayr, E. (1982). *The Growth of Biological Thought: Diversity, Evolution, and Inheritance*. Cambridge, MA: Belknap Press of Harvard University Press.

Mayr, E. and Provine, W.B. (1980). *The Evolutionary Synthesis: Perspectives on the Unification of Biology*. Cambridge, MA: Harvard University Press.

McGregor, A.P., Orgogozo, V., Delon, I., Zanet, J., Srinivasan, D.G., Payre, F., and Stern, D.L. (2007). Morphological evolution through multiple cis-regulatory mutations at a single gene. *Nature* 448, 587–90.

Metscher, B.D. and Ahlberg, P.E. (1999). Zebrafish in context: uses of a laboratory model in comparative studies. *Developmental Biology* 210, 1–14.

Meyer, A. (1926). *Logik der Morphologie im Rahmen einer Logik der gesamten Biologie*. Berlin: Springer.

Milinkovitch, M.C. and Tzika, A. (2007). Escaping the mouse trap: the selection of new Evo-Devo model species. *J Exp Zoolog B Mol Dev Evol* 308, 337–46.

Müller, G.B. (1989). Ancestral patterns in bird development. *J Evol Biol* 2, 31–47.

(2005). Evolutionary developmental biology. In F. Wuketits and F. Ayala (eds.), *Handbook of Evolution*. San Diego: Wiley, vol. 2, pp. 87–115.

(2007a). Evo-devo: extending the evolutionary synthesis. *Nature Reviews Genetics* 8, 943–9.

(2007b). Six memos for EvoDevo. In M.D. Laubichler and J. Maienschein (eds.), *From Embryology to Evo Devo: A History of Developmental Evolution*. Cambridge, MA: MIT Press, pp. 499–524.

Müller, G.B. and Newman, S.A. (2003). *Origination of Organismal Form: Beyond the Gene in Developmental and Evolutionary Biology*. Cambridge, MA: MIT Press.

Müller, G.B. and Newman, S.A. (2005a). Editorial: evolutionary innovation and morphological novelty. *J Exp Zoolog B Mol Dev Evol* 304, 485–6.
(2005b). The innovation triad: an EvoDevo agenda. *J Exp Zoolog B Mol Dev Evol* 304, 487–503.
Naef, A. (1919). *Idealistische Morphologie und Phylogenetik (Zur Methodik der systematischen Morphologie)*. Jena: Gustav Fischer.
Neff, M.W., and Rine, J. (2006). A fetching model system. *Cell* 124, 229–31.
Negre, B., Casillas, S., Suzanne, M., Sánchez-Herrero, E., Akam, M., Nefedov, M., Barbadilla, A., de Jong, P., and Ruiz, A. (2005). Conservation of regulatory sequences and gene expression patterns in the disintegrating *Drosophila Hox* gene complex. *Genome Res* 15, 692–700.
Nijhout, H.F. (2003). Development and evolution of adaptive polyphenisms. *Evol Dev* 5, 9–18.
Nordenskiöld, E. (1967). *Die Geschichte der Biologie. Ein Überblick*. Wiesbaden: M. Sändig.
Nyhart, L.K. (1995). *Biology Takes Form: Animal Morphology and the German Universities, 1800–1900*. University of Chicago Press.
(2002). Learning from history: morphology's challenges in Germany ca. 1900. *J Morphol* 252, 2–14.
Orgogozo, V., Broman, K.W., and Stern, D.L. (2006). High-resolution quantitative trait locus mapping reveals sign epistasis controlling ovariole number between two *Drosophila* species. *Genetics* 173, 197–205.
Page, R.E., Jr. (1997). The evolution of insect societies. *Endeavour* 21, 114–20.
Page, R.E., Jr. and Amdam, G.V. (2007). The making of a social insect: developmental architectures of social design. *Bioessays* 29, 334–43.
Peterson, K.J., Cameron, R.A., and Davidson, E.H. (2000). Bilaterian origins: significance of new experimental observations. *Dev Biol* 219, 1–17.
Prum, R.O. (2005). Evolution of the morphological innovations of feathers. *J Exp Zoolog B Mol Dev Evol* 304, 570–9.
Raff, R.A. and Wray, G. (1989). Heterochrony: developmental mechanisms and evolutionary results. *J Evol Biol* 2, 409–34.
Richards, R.J. (1992). *The Meaning of Evolution: The Morphological Construction and Ideological Reconstruction of Darwin's Theory*. University of Chicago Press.
(2002). *The Romantic Conception of Life: Science and Philosophy in the Age of Goethe*. University of Chicago Press.
Riedl, R. (1975). *Die Ordnung des Lebendigen: Systembedingungen d. Evolution*. Hamburg and Berlin: Parey.
Roger, J. and Williams, L.P. (1997). *Buffon: A Life in Natural History*. Ithaca, NY: Cornell University Press.
Rudwick, M.J.S. and Cuvier, G. (1997). *Georges Cuvier, Fossil Bones, and Geological Catastrophes: New Translations and Interpretations of the Primary Texts*. University of Chicago Press.
Russell, E.S. (1916). *Form and Function: A Contribution to the History of Animal Morphology*. London: J. Murray.

Salazar-Ciudad, I. (2006). Developmental constraints vs. variational properties: how pattern formation can help to understand evolution and development. *J Exp Zoolog B Mol Dev Evol* 306, 107–25.
Santini, S., Boore, J.L., and Meyer, A. (2003). Evolutionary conservation of regulatory elements in vertebrate *Hox* gene clusters. *Genome Res* 13, 1111–22.
Sapp, J. (2003). *Genesis: The Evolution of Biology*. New York and Oxford: Oxford University Press.
Schlosser, G. and Wagner, G.P. (2004). *Modularity in Development and Evolution*. University of Chicago Press.
Shi, W., Levine, M., and Davidson, B. (2005). Unraveling genomic regulatory networks in the simple chordate, *Ciona intestinalis*. *Genome Res* 15, 1668–74.
Stathopoulos, A. and Levine, M. (2002). Dorsal gradient networks in the *Drosophila* embryo. *Dev Biol* 246, 57–67.
(2004). Whole-genome analysis of *Drosophila* gastrulation. *Curr Opin Genet Dev* 14, 477–84.
Stern, D.L. (2003). The Hox gene *Ultrabithorax* modulates the shape and size of the third leg of *Drosophila* by influencing diverse mechanisms. *Dev Biol* 256, 355–66.
(2006). Perspective: developmental biology – morphing into shape. *Science* 313, 50–1.
(2007). The developmental genetics of microevolution. *Novartis Found Symp* 284, 191–200; discussion 200–6.
Sucena, E., Delon, I., Jones, I., Payre, F., and Stern, D.L. (2003). Regulatory evolution of shavenbaby/ovo underlies multiple cases of morphological parallelism. *Nature* 424, 935–8.
Thompson, D.A.W. (1917). *On Growth and Form*. Cambridge University Press.
Toth, A.L. and Robinson, G.E. (2007). Evo-devo and the evolution of social behavior. *Trends Genet* 23, 334–41.
Troll, W. (1928). *Organisation und Gestalt im Bereich der Blüte*. Berlin: J. Springer.
Wagner, G.P. (2000). What is the promise of developmental evolution? Part I: why is developmental biology necessary to explain evolutionary innovations? *J Exp Zool* 288, 95–8.
(2001). What is the promise of developmental evolution? Part II: a causal explanation of evolutionary innovations may be impossible. *J Exp Zool* 291, 305–9.
Wagner, G.P., Chiu, C.-H., and Laubichler, M.D. (2000). Developmental evolution as a mechanistic science: the inference from developmental mechanisms to evolutionary processes. *American Zoologist* 40, 819–31.
Wagner, G.P. and Larsson, H.C. (2003). What is the promise of developmental evolution? Part III: the crucible of developmental evolution. *J Exp Zoolog B Mol Dev Evol* 300, 1–4.
West-Eberhard, M.J. (2003). *Developmental Plasticity and Evolution*. Oxford and New York: Oxford University Press.

Wilson, E.B. (1925). *The Cell in Development and Heredity*. New York: The Macmillan Company.

Wimsatt, W.C. (2007). Echoes of Haeckel? Reentrenching development in evolution. In M.D. Laubichler and J. Maienschein (eds.), *From Embryology to Evo-Devo: A History of Developmental Evolution*. Cambridge, MA: MIT Press, pp. 309–56.

3

Deducing plant function from organic form: challenges and pitfalls

KARL J. NIKLAS

"All we know in advance when we construct a machine is the function it will have, not its structure or form."

"We cannot use general principles to sum up historical reality any more than we can use the principles of thermodynamics to construct a Rolls-Royce." Eigen and Winkler (1983)

INTRODUCTION

A common thread throughout the history of comparative morphology is the debate over the primacy of form versus function (Russell 1916). Does form follow function, or does function follow form? This "antithetic" argument appears to have an almost universal appeal – it recurs in the analysis of art, literature, and architecture, as well as biology, where explicit analogies with "design" are often made. Historically, some of the leading morphologists of their day had strong and divided opinions. Others saw the antithetic argument as a non-issue. Geoffroy believed that form determines function. For him, form had supremacy. In contrast, Cuvier, like Aristotle, argued that functional and structural "harmony" is essential to ecological success, without which a species would not exist. Because species obviously do exist, it logically follows that this "harmony" exists. Cuvier therefore argued strongly that structure and function had to be treated as one and inseparable. Arber too felt that the contrast between form and function ceases to exist once the word "form" is given its full biological content. However, for her, the opposition of the two concepts was the simple, albeit insidious result of a mistaken analogy between human artifacts and living things.

Certainly, it is possible to concentrate on the study of one to the exclusion of the other. Yet the antithetic dichotomy is not symmetrical. It is possible to study and use the diversity of organic form without being concerned about function. For example, the taxonomist can classify the diversity of form without knowing anything about what form "does." But the reverse is not true. It is not possible to study organic function without detailed knowledge of its attending form and internal organization.

How then is organic function studied? Much of the published research in functional morphology follows the general guidelines of the "paradigm method" suggested by Rudwick (1964, 1968), which involves four steps: (1) a function is first proposed for a structure or set of structures; (2) it is then structurally "transcribed" into engineering terms to identify an artifact that provides the optimal efficiency in the performance of the posited function; (3) the resulting engineering "paradigm" is then compared quantitatively to the actual structure or set of structures the organism possesses; and (4) the extent to which the paradigm and the structure (or set of structures) comply with one another is taken as evidence that the structure has the ability to perform the hypothesized function. Even though Rudwick's paradigm method was originally proposed to deal with fossil structures whose functions cannot be observed directly, his approach differs little or not at all from how form–function relationships of extant organisms are typically examined.

Several authors have criticized Rudwick's methodology (e.g. Gould 1970; Gould and Lewontin 1979; Grant 1972). Four criticisms are particularly sharp: (1) it is possible that the paradigm method will never demonstrate that a particular structure is non-functional or maladaptive; (2) it is possible to select the wrong paradigm, yet still find a correspondence between it and the structure being examined; (3) a structure may have several functions and thus not fit any one paradigm very closely; and (4) historical constraints imposed by body plans might limit the extent to which a structure fits a particular paradigm.

Arguably, all of these criticisms revolve around the a priori supposition that a structure (or set of structures) has but a *single* function, such that it is functionally *optimal* or nearly so. This supposition is central to many of the debates concerning twentieth-century evolutionary morphology, whose major focus was to understand the history of natural selection in reference to the evolution of organic structures (see Bock 1980; Cracraft 1981; Gans 1974). The use of optimization models in morphology grew out of the post-Darwinian view that organisms are constrained more by the demands placed on them by their environment than by their bauplan and intrinsic organizational hierarchy (see Gould 1980). Thus, for example, Rosen (1967: 68) claimed that "selection pressure is the major determinant of structure." If it is true that the effect of selection is an optimization of structure, then phyletic transformation sequences should be governed largely by the principles of optimality and much less so by phyletic legacy.

Curiously, this post-Darwinian view neglects the fact that selection can only occur among structural variants that exist (which are as much a consequence of random genomic changes as they are the result of prior selection). Nevertheless, many morphologists emphasized either the supremacy of "optimality" or that of "phyletic legacy." For example, D'Arcy Thompson rejected the historical approach to the study of form. Although his coordinated grid-form transformations can be interpreted in the context of evolutionary change, Thompson (1917) searched for purely mechanical or physical factors capable of producing or explaining form. This approach, which emphasizes the equilibrium concept of the organism and its environment – one in which history can be largely ignored as a component in the explication of organic form – continued to dominate much of twentieth-century biomechanical inquiry (see Niklas 1992; Vogel 1981; Wainwright *et al.* 1976).

Certainly, a major criticism of the use of optimality models is that they may fail to characterize accurately a system in which intrinsic historical factors have been important determinants of form. Indeed, this pitfall is highlighted nowhere better than when we juxtapose the basic tenets of biomechanics with those of its conceptual antecedent, engineering. Although every engineered artifact is designed to function with a priori knowledge of its functional obligations and working environment, the design and level of performance of every mechanical device are intrinsically limited by prior technological practices and theoretical insights. The availability of building materials and the ability to use them wisely impose a "historical legacy" on every engineered structure every bit as real as the constraints placed on organic form exacted by an organism's evolutionary history.

The foregoing issues have guided the agenda of this chapter, which is to explore the implications of following a strict optimality approach to understanding plant form–function relationships and to explore the consequences of historical events on the interpretation of these relationships. This agenda is pursued in two very different, albeit interrelated ways – first, by means of computer simulations specifically tailored to examine early vascular land plant evolution; and, second, by examining the biomass partitioning patterns of diverse extant plant species.

Computer simulations of ancient land plant evolution have the advantage of providing unambiguous renderings of the optimization process; early land plant evolution is emphasized because each of the vegetative body parts of the most ancient tracheophytes was multifunctional. Indeed, even with the evolution of comparatively sharp developmental distinctions among foliage leaves, self-supporting stems, and subterranean roots, each organ-type in the organographic triumphant of the stereotypical vascular sporophyte continues to be multifunctional – for example, foliage leaves are mechanically, hydraulically, and photosynthetically integrated functional units. Thus, the form–function relationships for plants are more often under selection against the inefficient performance of two or more functional tasks, rather than just one.

The phrase "biomass partitioning patterns" refers to the manner in which total body dry mass is divided among the three organ-types of the vegetative body at the level of modern-day individual plants (Niklas 2004a; Niklas and Enquist 2001, 2002). The scaling of these patterns is relevant to the agenda of this chapter, because, if statistically indistinguishable scaling relationships occur across phyletically very different plant lineages, they are likely to be the result of convergent (adaptive) evolution resulting in morphologically similar structures that are functionally equivalent. If, on the other hand, statistically very different patterns exist among evolutionarily different plant lineages, they likely reflect developmental constraints, thus highlighting the importance of phyletic legacy in the evolution of form–function relationships.

SIMULATIONS OF EARLY LAND PLANT EVOLUTION

This section is organized in the following way: a "morphospace" for all mathematically conceivable early vascular land plants is constructed; the variants existing in this space are evaluated quantitatively for how well they perform one or more of four basic biological functions; early land plant evolution is simulated by specifying the functions that must be performed and subsequently by locating successively more optimal variants; and the results of these simulations are discussed in the context of allometric evidence that leaf-, stem-, and root-like plant organs comprise "functionally equivalent" natural categories that trespass the phylogenetic boundaries of the polyphyletic algae and the monophyletic embryophytes.

The early vascular plant morphospace

The stereotypical tracheophyte sporophyte is composed of cylindrical, branched, axial elements. Among modern-day vascular plants, these elements are stems. However, those of the most ancient vascular plants cannot be so designated, because the anatomical and developmental distinctions among stems, leaves, and roots had yet to evolve (fig 3.1). Therefore, the morphospace for the most ancient vascular plants is comparatively simple to construct mathematically. Only a few parameters are required to construct virtually any phenotypic variant. These are the bifurcation angle f, the rotation angle g, the probability of apical bifurcation *P*, and the length *L* and diameter *D* of the axial elements (fig 3.2). Mathematically, $P=8p_{n-(k+1)} / (N+k)$, where $p_{n-(k+1)}$ is the probability of terminating branching at the next generated (higher) level of branching, and $N+k$ designates the previously generated level of branching (see Niklas 1994, 1997a, b, c; Niklas and Kerchner 1984). An additional parameterization is required to specify the manner in which *L* and *D* vary numerically from the base to the terminal axial element of each variant. For simplicity, *L* and *D* are specified to vary as a simple function of *P* such that the largest axial elements are at the base and the smallest elements subtend the most distal apical meristems. Assuming that the bulk tissue density r is uniform throughout the plant body, the mass *M* of any axial element is given by the formula π r $(D^2/4)$ *L*. With the addition of these few parameters, the location, size, and weight of each axial element in a variant are mathematically defined such that the display of all elements with respect to sunlight and the mechanical forces acting at the base of the variant are known precisely. Assuming that all terminal apical meristems convert into sporangia that freely disperse their spores by wind currents, the location and number of all sporangia are also specified.

The simplest morphospace consists of variants that are iso-dichotomous – that is, the bifurcation and rotation angles are "symmetrical" with respect to the longitudinal axis of the axial elements subtending each pair of elements (see fig 3.2). This morphospace is

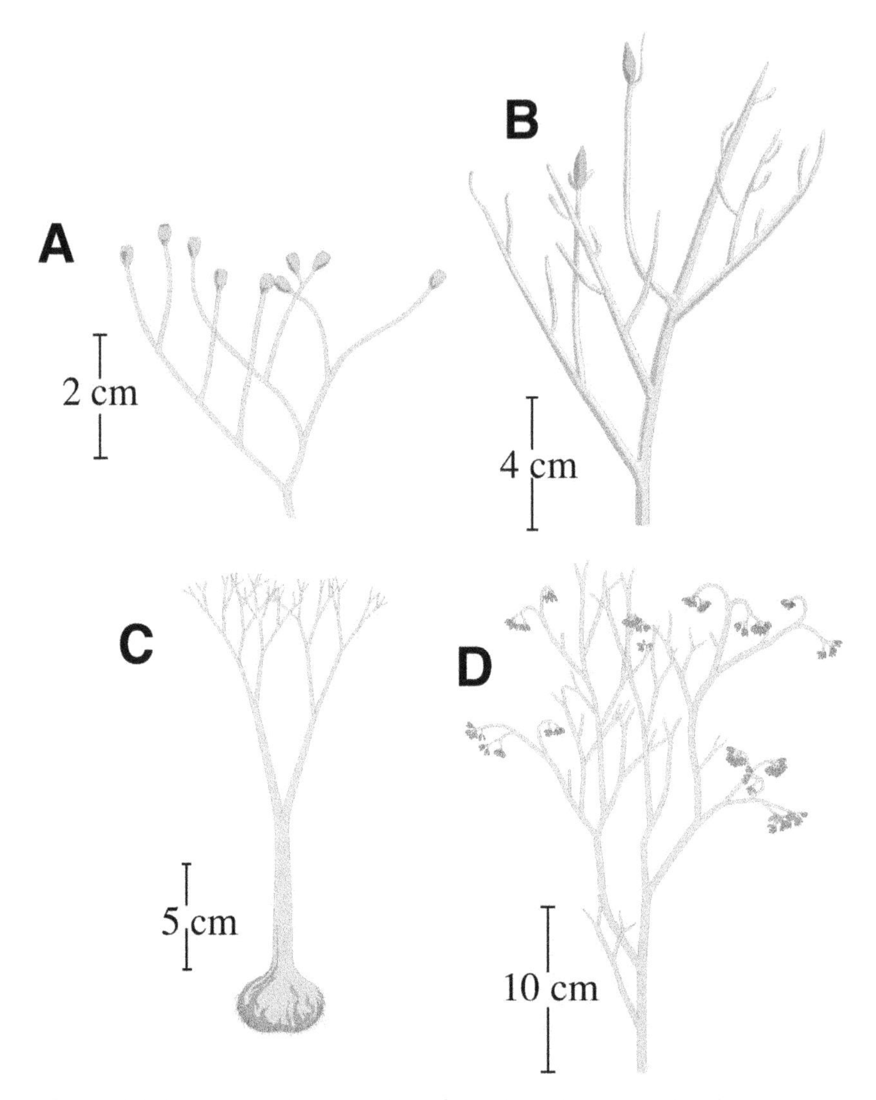

Figure 3.1 Representative morphologies of ancient vascular land plants. A. *Steganotheca striata* (redrawn from Edwards 1970). B. *Rhynia gwynne-vaughanii* (redrawn from Edwards 1980). C. *Horneophyton lignieri* (redrawn from Eggert 1974). D. *Psilophyton dapsile* (redrawn from Andrews *et al.* 1977).

a Cartesian space defined by the numerical values of γ, φ, and P. A multidimensional morphospace is required to cope with aniso-dichotomous variants, because two bifurcation angles, two rotation angles, and two probabilities of branching are required to construct these

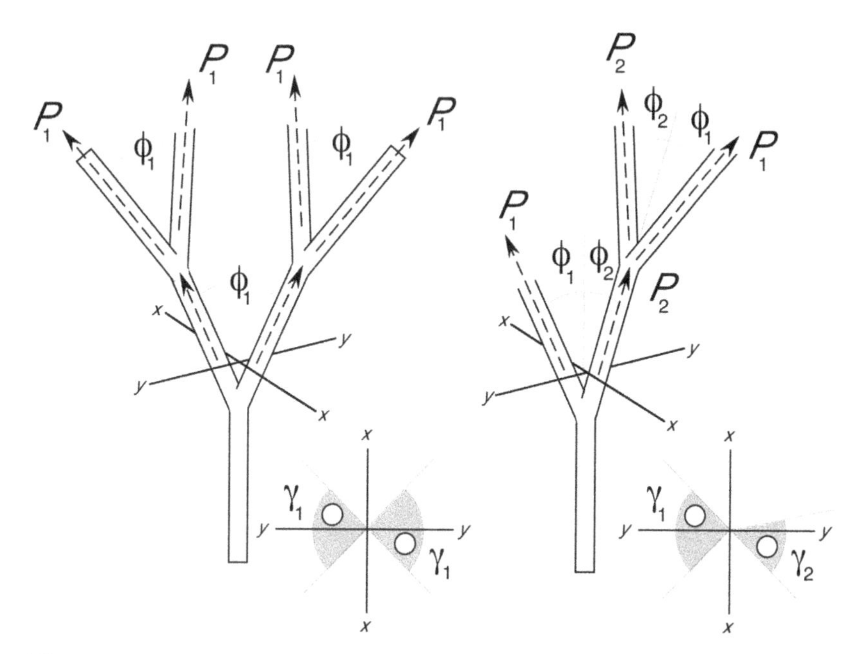

Figure 3.2 Mathematical variables required to construct iso- and aniso-bifurcating (left and right "stick" drawings, respectively) morphological variants of ancient vascular land plants. See text for details.

morphologies (see fig 3.2). Mathematically, the iso-dichotomous morphospace contains all variants for which $\varphi_1 = \varphi_2$ and $P_1 = P_2$. The aniso-dichotomous morphospace contains all morphologies for which $\varphi_1 \neq \varphi_2$, $\gamma_1 \neq \gamma_2$, and $P_1 \neq P_2$. In addition, the two axial elements subtended by each axis within each variant have different L and D, because each element has a different P-value (see Niklas 1994, 1997a, b, c; Niklas and Kerchner 1984).

Both morphospaces are constructed by independently varying each of the aforementioned variables. The spatial ordering of morphological variants is predetermined by assigning ascending numerical values to each of the variables – for example, the bifurcation angle varies in 1° increments; the probability of branching varies in 0.01 increments.

Relative fitness and fitness landscapes

The ability of each variant to perform one or more of four functions or "tasks" is easily evaluated using basic physics or engineering principles

(Niklas 1992). When the performance level of each variant is divided by the performance of the most efficient variant, the "relative fitness" of each variant is specified. It is assumed that each task contributes equally and independently to overall fitness such that the fitness of a particular variant, designated by h, is given by the formula $h = \tilde{A}^{1/N}$, where $\tilde{A}$ is the product of the performance levels of the number of tasks performed N. For example, if the numerical performance of two tasks is designated by the quantities A and B, the relative fitness of a particular variant is given by $h = (AB/\mathbf{AB})^{1/2}$, where $\boldsymbol{A}$ and $\boldsymbol{B}$ are the performance levels of the most efficient variant in the particular landscape.

As noted, all of the fitness landscapes are predicated on the performance of one or more of four tasks assumed to dictate survival and reproduction: light interception, maintaining mechanical stability, dispersing spores, and conserving water. These tasks can be considered in isolation or in sets of two or more tasks performed simultaneously. Thus, there are four 1-task landscapes, six 2-task landscapes, four 3-task landscapes, and one 4-task landscape. The methods for quantifying the performance of these tasks are provided in detail elsewhere (Niklas 1994, 1997b, c; Niklas and Kerchner 1984). However, each is reviewed briefly here.

Light interception is quantified for each morphological variant by integrating the area under the curve generated by plotting the projected surface S_p divided by the total surface area S of the variant as a function of the solar angle θ for the values $1° \leq \theta \leq 180°$, using an algorithm that accounts for the self-shading of axial elements (see Niklas and Kerchner 1984). Mechanical stability is calculated by computing the total bending moment on the basal axial element. Noting that a bending moment is the product of a bending force (weight W) and the length of a lever arm l, the bending moment of each distal axis in a branching architecture m_i is a function of $W = \pi D^2 L/4\mathrm{r}$, where D is diameter and r is the bulk tissue density (assumed to equal 1,000 kg/m^3 for all axes) and its bifurcation and rotation angles (which specify orientation and thus the lever arm length). The total bending moment at the base of any variant is a complex quantity to compute because it is the sum of the bending moments of all axial elements and because these elements are joined together and oriented differently as a function of the level of branching. The total bending moment must be computed numerically with the aid of a computer (see Niklas 1997a, b, c).

The ability of each variant to conserve water is taken as a simple linear function of total plant surface area – that is, it is assumed that all surface areas have equivalent rates of evapotranspiration. Finally, spore dispersal is assumed to rely exclusively on wind. A simple ballistic model is used, which assumes that all sporangia are equivalent in size and spore number. This model is given by the formula $x=HU/T$, where x is maximum distance of spore dispersal, H is plant height, U is ambient wind speed, and T is the spore terminal settling velocity (see Okubo and Levin 1989). Plant height is a function of the number, length, and orientation of all axes comprising a particular morphology; the ambient wind speed depends on plant height (since wind speeds typically decrease exponentially toward ground-level; and the terminal settling velocity is taken as 0.15 m/s (the average velocity of *Lycopodium* spores, which has become something of a "standard measure"). These assumptions and specifications obtain the scaling relationship $x \mu H^2$ across all morphologies – that is, the spore-dispersal range is, on average, proportional to the square of plant height.

Adaptive evolutionary walks

The fifteen fitness landscapes can be used to simulate "adaptive evolutionary walks" by means of a search algorithm that locates successively more efficient morphological variants defined on the basis of performing one or more designated tasks. Each walk begins at the same location in the morphospace. This location corresponds to the morphology of the most ancient vascular plants, such as *Cooksonia* and *Steganotheca* (see Taylor and Taylor 1993), which is located in the iso-dichotomous morphospace (see fig 3.1A). The search algorithm evaluates the relative fitness of all neighboring variants surrounding this location using a specified fitness criterion (e.g., light interception, light interception/mechanical stability, or light interception/mechanical stability/conservation of water). If one or more neighboring variants have an equivalent or higher relative fitness, the "walk" proceeds to their locations in the morphospace. This process continues until the relative fitness of the morphologies in the last iteration of a search is higher than that of all surrounding variants in the "terminal" steps. By definition, these morphologies occupy "adaptive peaks" on the landscape.

Adaptive walks can proceed through "stable" or "unstable" landscapes. Respectively, these are landscapes in which the task or set of tasks defining relative fitness remain constant or are changed arbitrarily at any time during a walk to mimic a change in selection. Stable landscapes can be used to explore the consequences of optimizing the performance of one or more tasks; unstable landscapes can be used to explore the consequences of history on subsequent adaptive form–function relationships.

Maximization and optimization in stable landscapes

Recall that two criticisms leveled against Rudwick's methodology (e.g. Gould 1970; Gould and Lewontin 1979; Grant 1972) are (1) that it is possible to select the wrong paradigm, yet still find a correspondence between it and the structure being examined; and (2) that a structure may have several functions and thus not fit any one paradigm very closely.

These two concerns are interrelated at a very fundamental level because they draw attention to the difference between a "good" mechanical device and the "best" possible. From an engineering perspective, a very real difference exists between the ability to *maximize* the performance of a single mechanical task and the ability to *optimize* the performance of two or more tasks that must be performed simultaneously. An artifact capable of "maximizing" its performance has the most efficient engineered configuration that is theoretically possible. An artifact that "optimizes" its performance is an efficient configuration, but one that is designed in the very practical context of potentially unavoidable mechanical limitations.

This difference resonates with the results of computer simulations of early vascular land plant evolution. These simulations indicate that comparatively few hypothetical morphologies are capable of maximizing the performance of any one of the four tasks required for plant growth, survival and reproductive success (fig 3.3), whereas the number of variants capable of optimizing the performance of two or more tasks simultaneously increases as the number of tasks increases (figs. 3.4–3.6). Likewise, the comparatively few morphological variants capable of maximizing the performance of any single task have the highest relative fitness, whereas the fitness of multitasking morphologies decreases as the number of tasks performed simultaneously increases (table 3.1).

Turning to specifics, the few hypothetical morphologies capable of maximizing the performance of any one of the four biological tasks used to quantify relative fitness are, for the most part, comparatively simple in general appearance (fig 3.3). The simplest of these are those that maximize the capacity to conserve water as gauged by minimizing their total surface area (fig 3.3A). These hypothetical phenotypic variants are

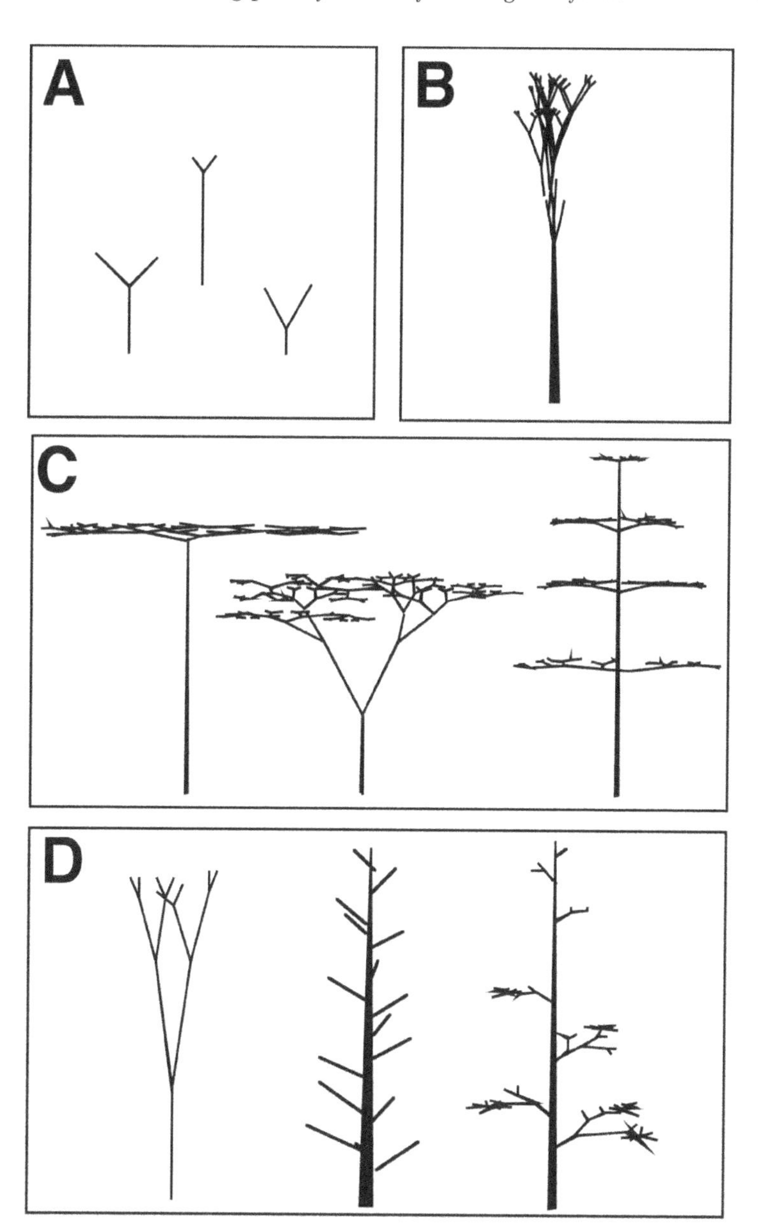

Figure 3.3 Morphological variants identified by "adaptive walks" through 1-task landscapes capable of maximizing the performance of one of four tasks. A. Water conservation. B. Spore dispersal. C. Light interception. D. Mechanical stability.

Table 3.1. *Relationships among the number of tasks used to quantify relative fitness, the number of morphological variants identified by "adaptive walks" that are capable of either maximizing or optimizing these tasks, and the relative fitness of these morphologies. n refers to the number of fitness landscape permutations*

Number of tasks defining fitness	Number of morphological variants (mean ± SE)	Relative fitness (mean±SE)
1 ($n=4$)	2.50 ± 1.0	35.3 ± 1.80
2 ($n=6$)	3.33 ± 1.6	11.6 ± 0.68
3 ($n=4$)	6.25 ± 0.5	7.5 ± 0.14
4 ($n=1$)	20	2.4

Y-shaped and similar in general appearance to the most ancient vascular plants – for example, *Cooksonia* and *Steganotheca* (fig 3.1A). Thus, water conservation may have been achieved early on in the evolutionary history of land colonization. However, from a mathematical perspective, these Y-shaped morphologies are trivial, because they are those used to initiate each adaptive walk, whether on single- or multiple-task landscapes.

In contrast, the most complex single-task morphologies are those that possess lateral branching systems confined in part or entirely to the horizontal plane and elevated on a single vertical "main" axis (fig 3.3C–D). Some of these morphologies are located in the iso-dichotomous morphospace, whereas others are found in the aniso-dichotomous morphospace for ancient tracheophytes. Because the latter are reached only by an extensive series of hypothetical morphological transformations, the adaptive walks reaching these morphologies are generally highly branched.

The morphologies capable of optimizing the performance of two or more tasks are typically far more diverse in general appearance than those maximizing the performance of a single task (figs. 3.4–3.6). Some of these variants are simple Y-shaped *Cooksonia*-like morphologies (fig 3.4A–C), whereas others are strikingly reminiscent of the fossil remains of some zosterophyllophytes, such as *Zosterophyllum*, an ancient group of vascular plants (fig 3.4D). As noted, the number of morphologies identified as equally efficient at performing two or more tasks simultaneously increases as the number of tasks increases (table 3.1). For example, on average, 3.3 morphological variants are reached by adaptive walks on the six 2-task landscapes (fig 3.4), whereas

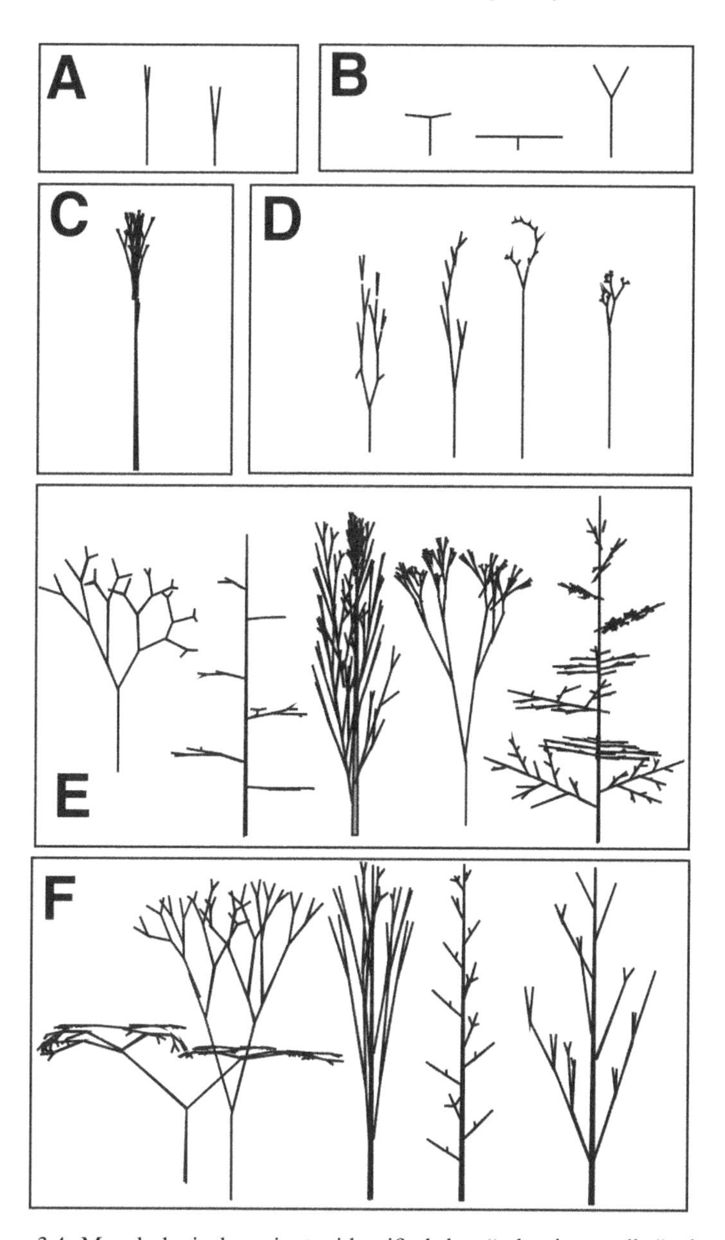

Figure 3.4 Morphological variants identified by "adaptive walks" through 2-task landscapes capable of optimizing the performance of two of four tasks. A. Mechanical stability and water conservation. B. Light interception and water conservation. C. Mechanical stability and spore dispersal. D. Spore dispersal and water conservation. E. Light interception and mechanical stability. F. Light interception and spore dispersal.

6.5 variants are reached by walks on the four 3-task landscapes (fig 3.5). When all four tasks are considered simultaneously, a total of 20 equally efficient morphologies is reached by an adaptive walk; many of these forms look strikingly like extant and extinct plants (fig 3.6).

Optimization in unstable (shifting) landscapes

Another criticism leveled against Rudwick's paradigm approach is that historical constraints might limit the extent to which a structure fits a particular paradigm. Environments can change, often dramatically, even over ecological timescales (10^2 to 10^4 yr). It is therefore naïve to believe that selection persistently acts on the performance of any one task or any particular combination of tasks, especially over timescales relevant to evolutionary or geological history (10^5 to 10^6 yr).

The consequences of "shifting" selection pressures can be modeled by initiating an adaptive walk on one landscape and subsequently changing the landscape one or more times, either randomly or in some proscribed sequence. In this way, the criteria used to quantify relative fitness change as a walk proceeds through the morphospace. Because the number of permutations of shifting fifteen fitness landscapes is astronomically large, it is reasonable to initiate all shifting adaptive walks on a 1-task fitness landscape and to subsequently shift to a 2-, 3-, and 4-task landscape.

Four pairs of these simulations are shown in figs. 3.7–3.10. In each pair, the simulation begins on the same landscape and comes to closure on the 4-task landscape. A recurring feature, which appears in all similar walks through unstable landscapes, is that adaptive walks entering the same fitness landscape locate different morphological optima depending on how relative fitness was defined in the 2- and 3-task landscapes. For example, the three optimal morphologies reached at the conclusion of each of the two adaptive walks entering the water-conservation landscape differ, even though the selection criteria were the same at the end of the walk (fig 3.7). These simulations indicate that the morphological variants with the highest relative fitness at any stage in adaptive evolution depend on the fittest morphologies defined in terms of prior selection regimes. Because these regimes confine the range of phenotypes that are available for the next round of selection, morphological "optimization" is historically contingent – that is, we should not expect adaptive evolution to achieve the best *conceivable* morphologies, but rather those that are "functionally superior" relative to what currently

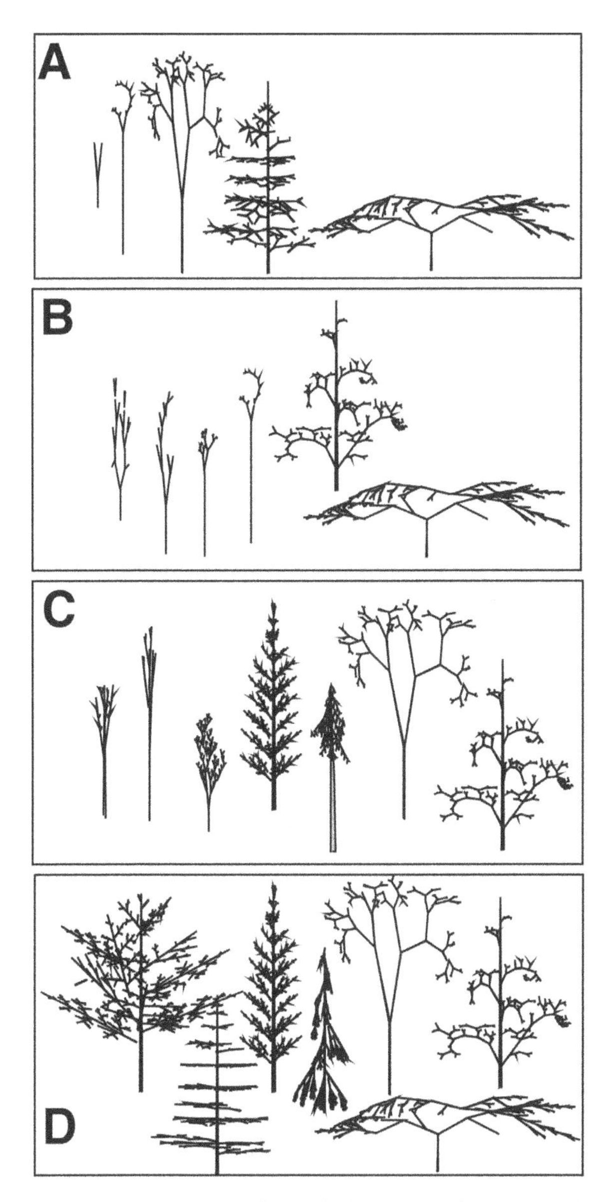

Figure 3.5 Morphological variants identified by "adaptive walks" through 3-task landscapes capable of optimizing the performance of three of four tasks. A. Light interception, mechanical stability, and water conservation. B. Light interception, spore dispersal, and water conservation. C. Mechanical stability, spore dispersal, and water conservation. D. Light interception, mechanical stability, and spore dispersal.

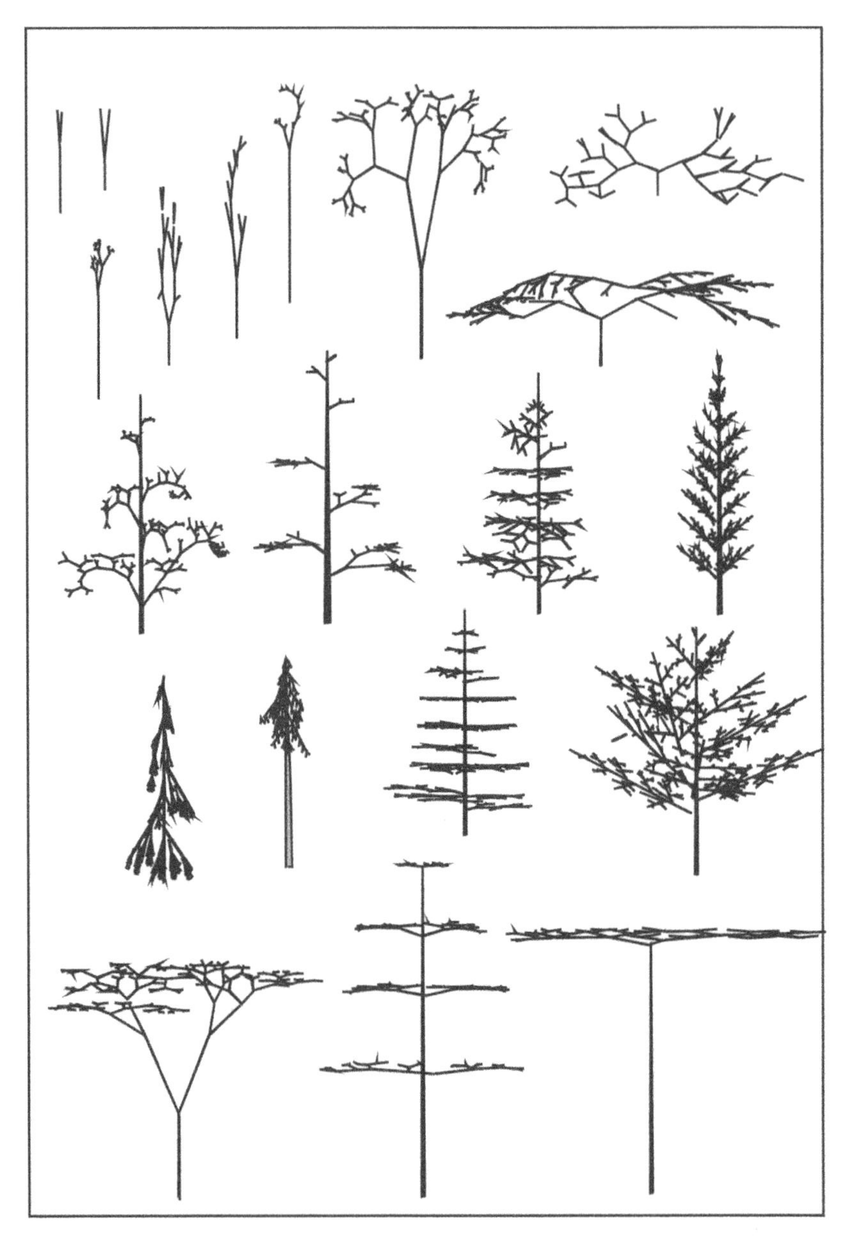

Figure 3.6 Morphological variants identified by "adaptive walks" through the 4-task landscape capable of optimizing the performance of all four tasks (mechanical stability, spore dispersal, water conservation, and light interception).

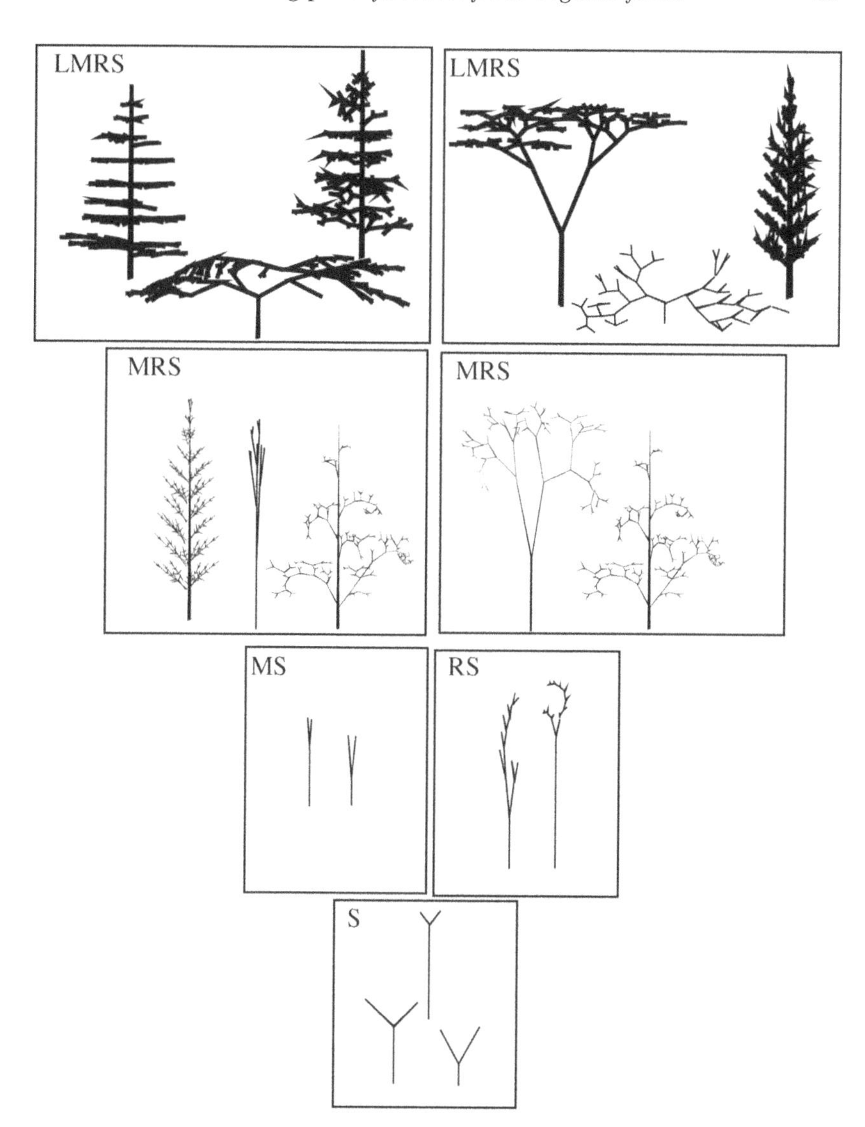

Figure 3.7 Morphological variants identified by two "adaptive walks" through two different sequences of shifting "adaptive landscapes," beginning with the 1-task landscape for water conservation (bottom) and ending with the 4-task landscape (top). Variants identified in the 4-task landscape by walks through different sequences of shifting landscapes are indicated in thicker lines (see figs. 3.8–3.10). Symbols used to identify tasks: L, light interception; M, mechanical stability; R, spore dispersal; and S, water conservation.

exists and what is developmentally available based on past selection regimes and their history. Importantly, however, many of the same morphological variants are reached by computer-simulated adaptive walks through very different prior selection regimes (see figs. 3.7–3.10) – that is, "adaptive convergence" is frequent despite previously different selection regimes.

THE SCALING OF EXTANT BIOMASS PARTITIONING PATTERNS

If form–function convergence is dictated more by "optimization" than by "historical contingency," phyletically very different lineages subjected to similar selection regimes should frequently converge on similar form–function relationships. This section is devoted to the exploration of whether similar form–function relationships (identified on the basis of biomass partitioning patterns) occur among radically different phyletic and ecological species-groups. It is organized in the following way. Prior work on the biomass partitioning patterns of modern-day seed plants is reviewed to construct a "null hypothesis" (i.e. a regression model with which to compare the scaling of the biomass partitioning patterns of non-seed plants and phyletically distinct species-groups). The data gathered from non-seed plants are juxtaposed with those from seed plants and statistical outliers are identified (i.e. data lying outside the 95 percent confidence intervals of the regression model for seed plant biomass partitioning patterns). Two hypotheses for explaining the extent to which the scaling of non-seed and seed plant biomass partitioning are evaluated in the context of the phyletic affiliations of statistical outliers (i.e. "development constraints" and "functional equivalence").

The scaling of seed plant (spermatophyte) patterns

The biomass partitioning pattern of living spermatophytes has been studied intensively (Enquist and Niklas 2001, 2002; Niklas, 2004a, 2005; Niklas and Enquist 2001, 2002). Across individuals differing in size but lacking the capacity for secondary tissues (or insufficiently old to accumulate these tissues), prior work has shown that leaf mass M_L scales, on average, isometrically with respect to both stem mass M_S and root mass M_R – that is, $M_L \propto M_S \propto M_R$. Therefore, the mass of all above-ground body parts ($M_A = M_L + M_S$) should scale isometrically or nearly so with respect to below-ground biomass ($M_A \propto M_B = M_R$) (Niklas 2004a, 2005;

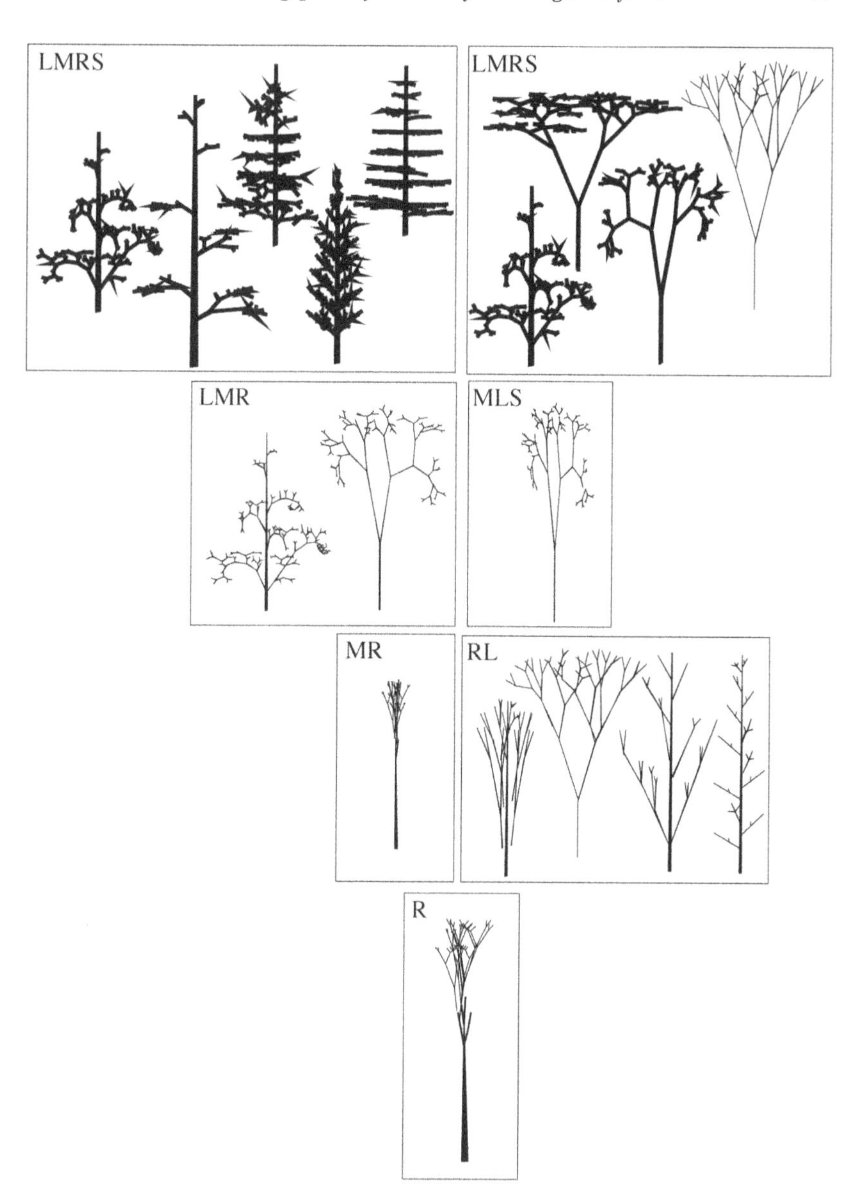

Figure 3.8 Morphological variants identified by two "adaptive walks" through two different sequences of shifting "adaptive landscapes," beginning with the single-task landscape for spore dispersal (bottom) and ending with the 4-task landscape (top). Variants identified in the 4-task landscape by walks through different sequences of shifting landscapes are indicated in thicker lines (see figs. 3.7, 3.9–3.10). Symbols used to identify tasks: L, light interception; M, mechanical stability; R, spore dispersal; and S, water conservation.

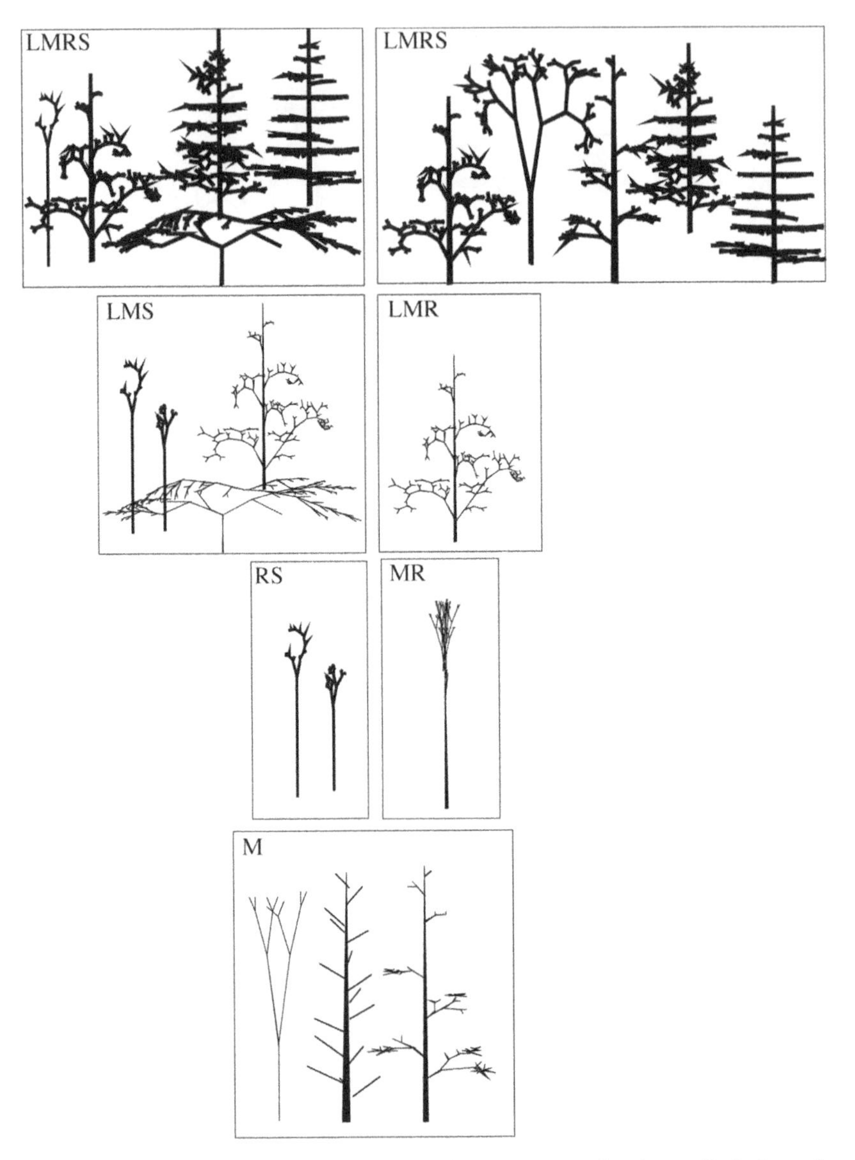

Figure 3.9 Morphological variants identified by two "adaptive walks" through two different sequences of shifting "adaptive landscapes," beginning with the single-task landscape for mechanical stability (bottom) and ending with the 4-task landscape (top). Variants identified in the 4-task landscape by walks through different sequences of shifting landscapes are indicated in thicker lines (see figs. 3.7–3.8, 3.10). Symbols used to identify tasks: L, light interception; M, mechanical stability; R, spore dispersal; and S, water conservation.

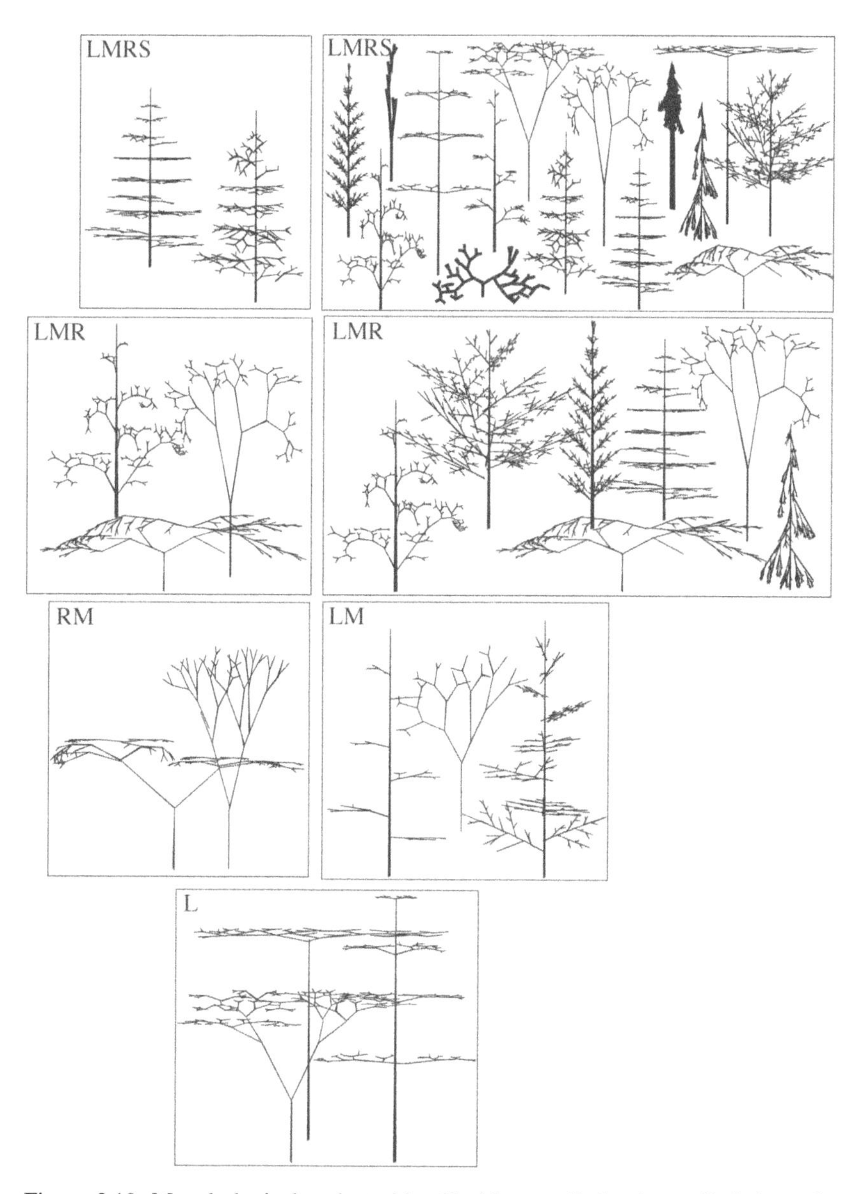

Figure 3.10 Morphological variants identified by two "adaptive walks" through two different sequences of shifting "adaptive landscapes," beginning with the single-task landscape for light interception (bottom) and ending with the 4-task landscape (top). Variants identified in the 4-task landscape by walks through different sequences of shifting landscapes are indicated in thicker lines (see figs. 3.7–3.9). Symbols used to identify tasks: L, light interception; M, mechanical stability; R, spore dispersal; and S, water conservation.

Niklas and Enquist 2001, 2002). In terms of the governing allometric constants (denoted successively with numerical subscripts), these proportionalities become $M_L = b_0\, M_S = b_1\, M_R$ such that $M_A = (1+1\,/\,b_0)b_1\, M_R$.

A fundamental question is whether these relationships hold true for taxonomically unrelated plants differing in their development and ecological requirements. If a single "canonical" partitioning pattern exists, it would provide strong circumstantial evidence that functional ("natural") organ-type categories exist and that these categories reflect evolution by natural selection on discrete form–function relationships.

Any comparison among the partitioning patterns of spermatophytes with those of other plant groups (such as the brown algae, bryophytes, or pteridophytes) necessarily treats taxa that either lack the capacity to produce secondary tissues or, if capable, produce extremely small quantities (see Bierhorst 1971). Therefore, the only legitimate comparisons that can be made between the biomass partitioning patterns previously reported for spermatophytes and those identified for non-spermatophytes is within the log-log linear phase represented by the proportional relationships $M_L \propto M_S \propto M_R$ and $M_A \propto M_R$. In this sense, the log-log linear phase of the spermatophyte biomass partitioning pattern can be considered the "null hypothesis." Taxa evidencing allometric deviations from this trend would provide evidence that no single partitioning pattern holds true across all plant lineages. Conversely, statistical concordance among the interspecific biomass partitioning patterns identified for non-spermatophyte and spermatophyte lineages would provide evidence that a single "canonical" pattern may exist.

One powerful statistical test for concordance between the scaling of spermatophyte and non-spermatophyte biomass relationships is the extent to which the 95 percent confidence intervals of allometric scaling exponent overlap numerically. This exponent is the slope of the log-log linear relationships among M_L, M_S, and M_R (or their analogs). The phyletic "out-groups" to the spermatophytes are the algae, mosses, and vascular non-seed plants. The specific tactic adopted is to (1) quantify the biomass partitioning pattern for all paired variables of interest across ecologically diverse spermatophytes represented by individuals lacking secondary tissues; (2) superimpose the same kinds of data obtained from algae, mosses, and pteridophytes; and (3) determine which (if any) non-spermatophyte taxa emerge as statistical outliers with respect to the allometry of spermatophytes (i.e. data that fall outside the 95 percent confidence intervals of spermatophyte scaling exponents).

Evidence for a canonical pattern

Across all paired variables of interest, the scaling exponents governing the biomass partitioning patterns of spermatophytes and non-spermatophyte lineages are statistically indistinguishable as gauged by their 95 percent confidence intervals. In the majority of cases, the scaling relationships among organ dry mass are isometric or nearly so – that is, the scaling exponents are numerically equal to one or nearly so (table 3.2). For example, across spermatophytes, leaf dry mass scales as the 0.90 (95% CI=0.88–0.92) power of dry stem mass; across non-spermatophyte streptophytes (= charophycean algae, mosses, and pteridophytes), M_L scales as the 0.91 (95% CI=0.70–1.13) power of M_S; and, across brown algal macrophytes, M_L (= blade dry mass) scales as the 0.83 (95% CI= 0.47–1.20) power of M_S (= stipe dry mass).

Table 3.2. *Summary statistics of reduced major axis (RMA) regression of log_{10}-transformed data for seed plant (spermatophytes) leaf, stem, and root dry mass (original units in kg; denoted by M_L, M_S, M_R, respectively), above- and below-ground body parts (denoted by M_A and $M_B = M_R$, respectively), and analogous data for non-spermatophyte land plants/charophycean algae and phaeophycean (brown algal) macrophytes. a_{RMA}, slope of regression curve (scaling exoponent); log b_{RMA}, Y–intercept of regression curve (allometric "constant")*

	log a_{RMA} (95% CI)	log b_{RMA} (95% CI)	r^2	n
Spermatophytes				
log M_L vs log M_S	0.90 (0.88, 0.92)	−0.19 (−0.25, −0.13)	0.91	862
log M_L vs log M_R	0.93 (0.91, 0.96)	−0.03 (−0.11, −0.06)	0.87	668
log M_S vs log M_R	1.01 (0.98, 1.04)	0.10 (−0.01, 0.20)	0.87	673
log M_A vs log M_B	1.04 (1.02, 1.06)	0.38 (0.31, 0.45)	0.88	1,223
Non-spermatophyte land plants/charophytes				
log M_L vs log M_S	0.91 (0.70, 1.13)	−0.05 (−0.83, 0.73)	0.86	16
log M_L vs log M_R	0.78 (0.51, 1.06)	−0.66 (−1.58, 0.25)	0.84	16
log M_S vs log M_R	0.96 (0.68, 1.24)	−0.26 (−1.22, 0.70)	0.89	16
log M_A vs log M_B	0.85 (0.64, 1.07)	−0.31 (−1.05, 0.43)	0.91	20
Phaeophycean macrophytes (all Laminariales)				
log M_L vs log M_S	0.83 (0.47, 1.20)	0.20 (−0.65, 1.06)	0.92	8
log M_L vs log M_R	0.87 (0.58, 1.16)	0.52 (−0.22, 1.26)	0.95	8
log M_S vs log M_R	1.04 (0.57, 1.51)	0.38 (−0.83, 1.58)	0.91	8
log M_A vs log M_B	0.89 (0.80, 1.03)	0.59 (0.32, 0.86)	0.99	8

However, these confidence intervals are numerically too broad to rigorously test whether statistical differences exist between the different species-groupings. A much more conservative approach (one that actually biases in favor of rejecting the notion that phyletically dissimilar species-groups share the same biomass partitioning patterns) is to examine the taxonomic affiliation of statistical outliers in the context of spermatophyte scaling relationships, thus drawing pointed attention to exceptions to the "rule."

Conservatively speaking, bivariate plots of all of the paired variables of interest reveal few, if any, brown algal statistical outliers, and those that do occur are outnumbered by embryophytes that are not seed plants (figs. 3.11–3.14). For example, five non-seed plant outliers (among which two are phaeophycean macrophytes) are observed for the scaling of leaf (or blade) mass with respect to stem (or stipe) mass (fig 3.11). Among the five outliers, three are pteridophytes – a fern (*Botrychium virginianum*), a lycopod (*Hupertzia lucidulum*, formerly known as *Lycopodium lucidulum*), and a horsetail (*Equisetum arvense*). Morphologically, these outliers should come as no surprise. Like all extant horsetails and the majority of lycopods, *E. arvense* and *H. lucidulum* have small ("microphyllous") leaves (Bierhorst 1971; Bold 1967; Gifford and Foster 1988). Those of *E. arvense* are so vestigial that they are non-photosynthetic, much as those found on the parasitic angiosperm *Monotropa uniflora*. In contrast, the "macrophyllous" leaves of *B. virginianum* are exceptionally large in comparison to their subterranean subtending stems. These differences in relative leaf size are sufficient to explain why three outliers are pteridophytes and why they respectively fall below or above the 95 percent confidence intervals of the spermatophyte M_L vs M_S scaling relationship (fig 3.11).

Likewise, the expansive indeterminate growth of the blades of *Laminaria* and *Saccorhiza* locate these two outliers above the 95 percent confidence intervals for the M_L vs. M_S bivariate plot (fig 3.11). It cannot escape attention that an aquatic habitat imposes little or no constraint on the amplification of surface area with respect to plant body volume, whereas plants that produce leafy organs with large surface areas in a terrestrial habitat run the risk of dehydration without benefit of a cuticle, stomata, and a hydraulic tissue capable of providing for the bulk flow of water to aerial organs (e.g. the fern *B. virginianum*).

Two additional observations are relevant to making comparisons across these or other phyletically or ecologically different species-groupings. First, although limited in number, all the data drawn from

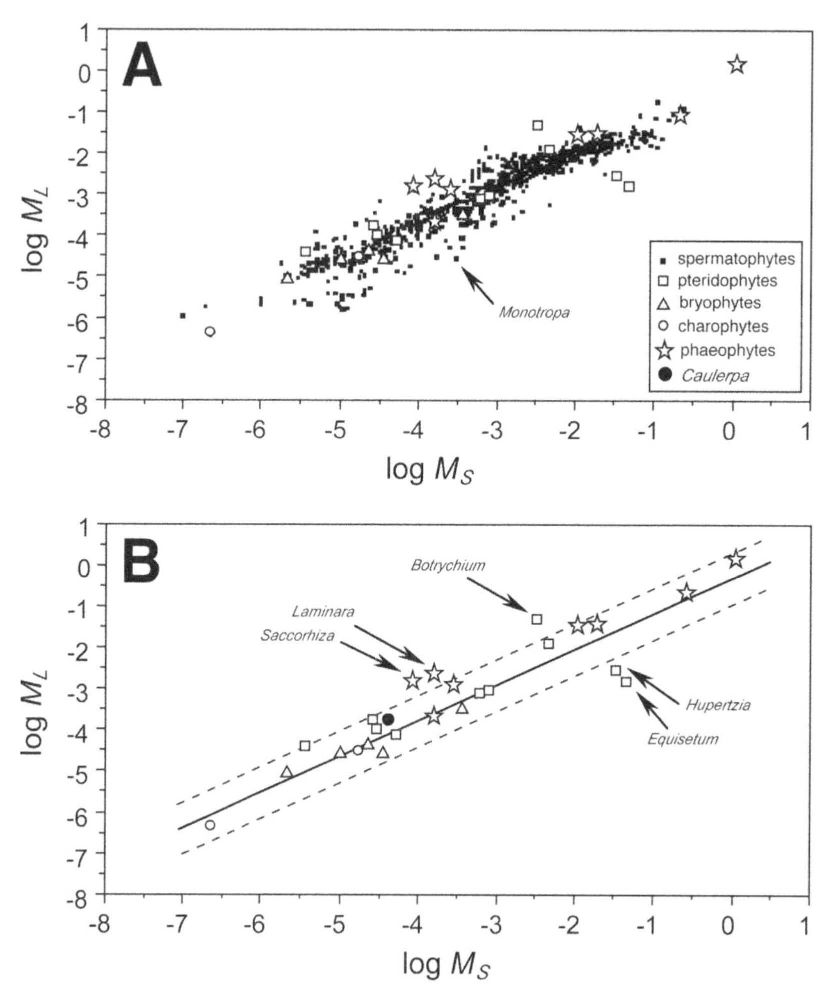

Figure 3.11 Bivariate log-log plots of standing leaf dry mass M_L versus standing stem dry mass M_S across individual seed plants (spermatophytes), pteridophytes, bryophytes, charophytes, marine brown algal macrophytes (phaeophytes), and the unicellular green alga *Caulerpa* (see insert for symbols). A. Bivariate plot for all data. B. Bivariate plot for data from non-seed plants. Six statistical outliers are indicated by arrows (data that exceed the 95 percent confidence intervals of the log-log regression curve; dashed and solid lines, respectively): *Monotropa* (parasitic flowering plant), *Laminaria* and *Saccorhiza* (brown algae), *Botrychium* (a fern), *Equisetum* (horsetail), and *Hupertzia* (a lycopod).

charophycean algae and from non-vascular land plants (i.e. mosses) plot directly on or very near the isometric RMA regression curves obtained exclusively using spermatophyte data; and, second, the data from

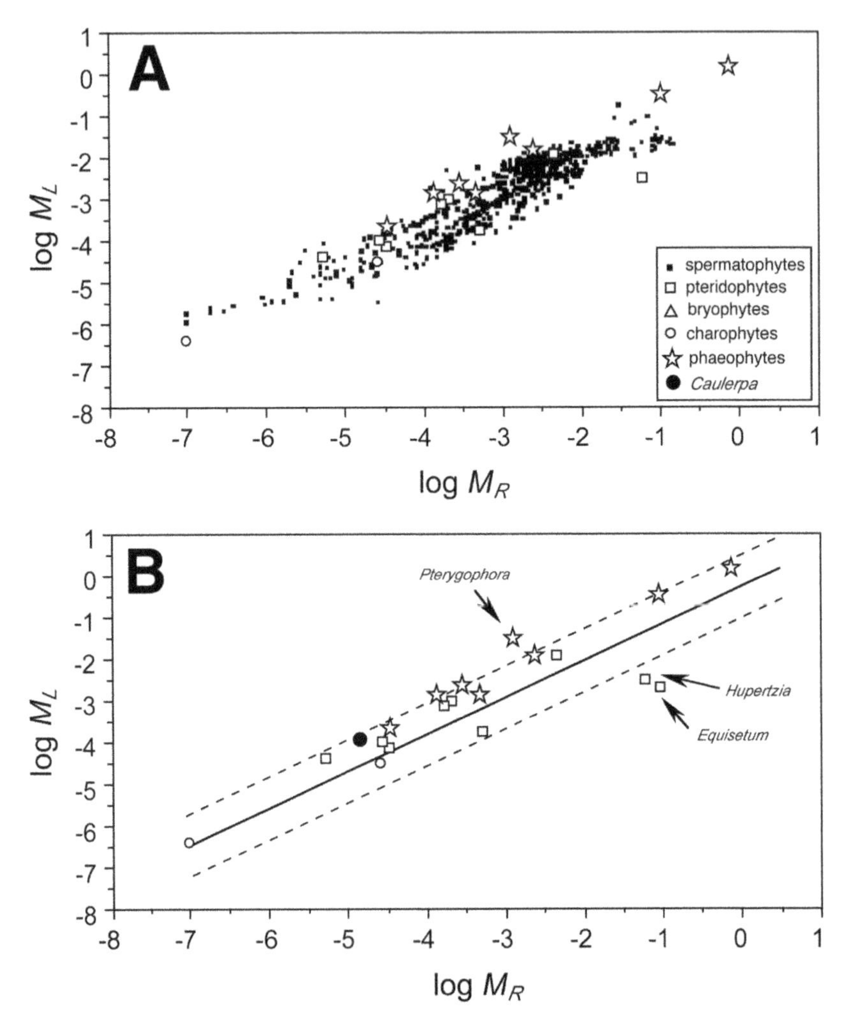

Figure 3.12 Bivariate log-log plots of standing leaf dry mass M_L versus standing root dry mass M_R across individual seed plants (spermatophytes), pteridophytes, bryophytes, charophytes, marine brown algal macrophytes (phaeophytes), and the unicellular green alga *Caulerpa* (see insert for symbols). A. Bivariate plot for all data. B. Bivariate plot for data from non-seed plants. Three statistical outliers are indicated by arrows (data that exceed the 95 percent confidence intervals of the log-log regression curve; dashed and solid lines, respectively): *Pterygophora* (brown algae), *Equisetum* (horsetail), and *Hupertzia* (a lycopod).

arguably the closest modern analog to the oldest known tracheophyte fossils (i.e. *Psilotum nudum*) fall directly on the RMA regression curve for the scaling of above- with respect to below-ground spermatophyte

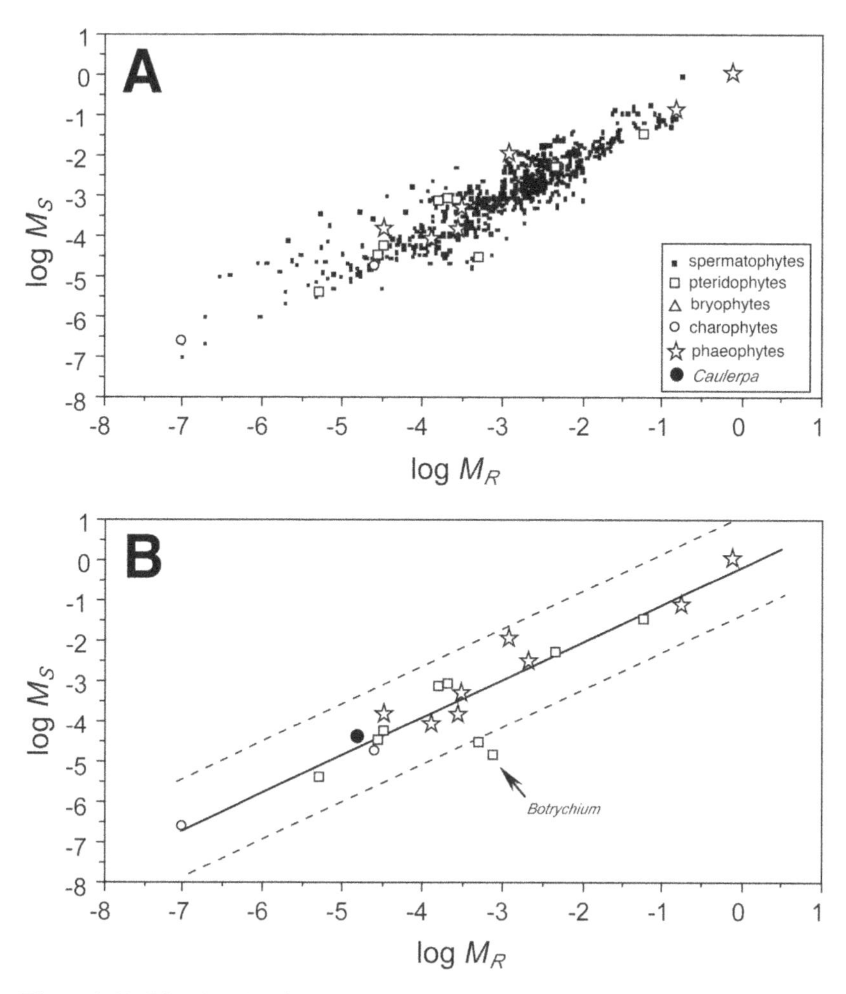

Figure 3.13 Bivariate log-log plots of standing stem dry mass M_S versus standing root dry mass M_R across individual seed plants (spermatophytes), pteridophytes, bryophytes, charophytes, marine brown algal macrophytes (phaeophytes), and the unicellular green alga *Caulerpa* (see insert for symbols). A. Bivariate plot for all data. B. Bivariate plot for data from non-seed plants. One statistical outlier is indicated by arrows (data that exceed the 95 percent confidence intervals of the log-log regression curve; dashed and solid lines, respectively): *Botrychium* (a fern).

biomass (fig 3.14). The statistical "invariance" of the biomass partitioning patterns observed across all streptophytes (= charophycean algae +embryophytes) is consistent with the fact that these plants constitute a monophyletic group and thus may reflect a deeply embedded

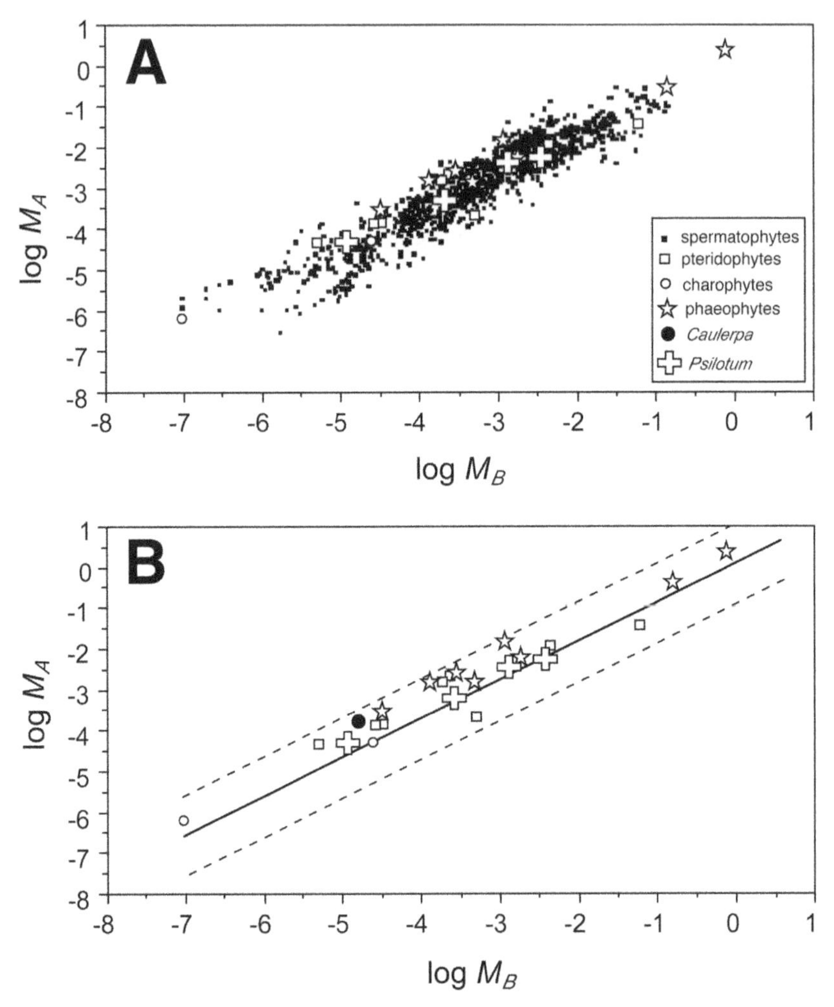

Figure 3.14 Bivariate log-log plots of standing above-ground dry mass M_A versus standing below-ground dry mass M_A across individual seed plants (spermatophytes), pteridophytes, bryophytes, charophytes, marine brown algal macrophytes (phaeophytes), the unicellular green alga *Caulerpa*, and the vascular land plant *Psilotum nudum* (see insert for symbols). A. Bivariate plot for all data. B. Bivariate plot for data from non-seed plants. No non-seed statistical outliers are evident (no data exceed the 95 percent confidence intervals of the log-log regression curve; dashed and solid lines, respectively).

developmental repertoire that transcends the genomic and ecological differences between the essentially aquatic haploid generation (that dominates the charophycean and bryophyte life cycles) and the largely terrestrial diploid generation (that dominates the tracheophyte life cycle). When viewed in this light, the allometry of non-spermatophyte streptophytes does not constitute an "out-group" comparison to that of spermatophyte allometric trends.

Two hypotheses

As noted, there are two possible interpretations for the statistical concordance among plant biomass partitioning patterns: (1) either developmental constraints act on the evolution of shared characters, or (2) leaf-, stem-, and root-like organ analogs are functional equivalent despite the developmental and ecological differences among lineages (Harvey and Pagel 1991; Niklas 1994). The "developmental constraint" hypothesis posits that natural selection acts on different body parts in opposing directions (gauged by biomass or some linear dimension), and that developmental synergistic interactions among these parts limit the extent to which one or more body parts can change in size evolutionarily. The "functional equivalence" hypothesis argues that particular body parts must change in size with respect to changes in the size of other body parts to maintain comparable functional levels of performance dictated by biophysically or physiologically invariant "rules."

Arguably, these two hypotheses are not mutually exclusive. Natural selection operating on the level of how well organs perform certain biological tasks can act indirectly on the developmental patterns that give rise to organ structure, shape, size, and so on. Put differently, if function-function variants are the objects of selection, then the developmental variations that give rise to them will also be the objects of selection. Certainly, organisms can obviate neither the laws of physics or chemistry nor the principles of engineering or mathematics. Thus, constraints on development have existed since the dawn of life. Nevertheless, the concepts of "developmental constraints" *sensu stricto* and "constraints on development" are different – the first posits that developmental repertoires are "internally" regulated and limited genomically; the second argues that the external environment limits which among possible developmental patterns persist in evolutionary time. This is not a semantic issue.

Yet, in the present case, the "developmental constraint" hypothesis *sensu stricto* can be rejected, because the scaling of biomass is remarkably concordant among phyletically unrelated and ecologically diverse species-groups. This is no more evident than for the allometry of phaeophycean macrophytes (here limited to members of the Laminariales), which unquestionably represent a line of plant evolution that is entirely independent of that of the algal ancestors of the charophycean algae and embryophytes (i.e. streptophiles). Likewise, all of the evidence in hand indicates that the development underwriting the blade, stipe, and holdfast construction of laminarian algae is radically different from that giving rise to tracheophyte leaves, stems, and roots (Anderson 2004; Graham and Wilcox 2000; Keeling 2004).

Specifically, the ancestors of the brown and the green algae trace their evolutionary origins back to independent primary endosymbiotic events in ancient Precambrian times (see Graham and Wilcox 2000; Niklas 2004b). Because the ancestral condition in each of these two lineages had to have a unicellular body plan, the evolution of the multicellular condition and how it variously achieves its organized growth in each lineage must have proceeded along dissimilar avenues of developmental innovation. Apical meristems and a parenchymatous tissue construction occur in both lineages, but in ways that can only be construed as adaptively convergent rather than homologous developmentally (except perhaps at the most basic level of transcription factors shared by all eukaryotic life forms).

Functional equivalence and its "latitude"

Despite the effort to reject the hypothesis, a single (isometric) biomass partitioning pattern appears to hold true across a broad phyletic spectrum of algae, bryophytes, pteridophytes, and spermatophytes. Perhaps most surprising is that, this pattern also holds true for the unicellular macroscopic alga *Caulerpa prolifera*, whose leaf-, rhizome (stem)- and root-like body construction continues to provide additional insight into the debate over whether "cells make organisms" or "organisms make cells" (see Kaplan 1992; Kaplan and Hagemann 1991). In addition, the data presented here appear to support the "functional equivalence" hypothesis rather than the "developmental constraint" hypothesis *sensu stricto*, because, regardless of their developmental or structural differences, what have been traditionally identified as functionally equivalent body parts seem to manifest the same biomass partitioning patterns. If the evidence is accepted at face value, then tracheophyte leaves, bryophyte phyllids, and algal macrophyte leaf-like structures must be construed to constitute a natural "functional" organ-category (as do stems-gametangiophores-internodal cells, or roots-rhizoids/protonema-holdfasts).

Arguably, the existence of natural functional organ-categories resonates reasonably well with the functional obligations that have traditionally been ascribed to each of the analogous body parts assigned to one of the three functional organ-categories. The foliose structures of marine green, red, or brown algal macrophytes intercept sunlight and exchange gasses with the fluid that surrounds them in physical and chemical ways that are not fundamentally dissimilar from those influencing the exchange of mass and energy between the air and the foliage leaves of tracheophytes. Likewise, despite differences in their hydraulic (and thus anatomical) obligations, roots and holdfasts both provide anchorage to a substrate, just as the aerial stems of vascular plants, the gametangiophores of bryophytes, and the stipes of many algae are nearly identical from an engineering perspective (Denny *et al.* 1997; Holbrook *et al.* 1991; Koehl and Wainwright 1977; Santelices 2004).

That the size-dependent (scaling) requirements of developmentally different structures (which appear to share the same or very similar functional obligations) are not truly "invariant" is evident from the inspection of the absolute rather than the proportional biomass relationships among leaf, stem, and root analogs. As noted, the scaling exponent describes the proportional relationship between two variables, whereas the *Y*-intercept is required to compute the absolute values of the variable of interest. table 3.2 shows that the *Y*-intercepts for some species-groups have numerically broad 95 percent confidence intervals, which indicates that these species-groups occupy a wide range of body part sizes.

This statistical "latitude" can be interpreted a posteriori as a reflection of the latitude permitted in the range of biomass occupied by each functional organ-category, which is otherwise confined by the operation of physical and chemical laws or processes dictating the functional performance of each organ-category. Metaphorically speaking, the data-scatter observed across all of the taxa identifies the "size corridors" through which plants have evolved as their size ranges expanded or contracted over geological timescales. That these corridors seem to have well-defined limits attests to the operation of these physical and chemical laws and processes. That some taxa "pushed" these limits and, in some cases, exceeded them attests to the ability of life to evolve innovative chemical or physical solutions to biotic and abiotic challenges.

CONCLUSIONS AND CLOSING REMARKS

Two conclusions can be drawn from the preceding analyses: "optimization" and "historical contingency" both play important roles during simulations of early land plant form–function evolution, yet optimization (as a result of selection driven by invariant physical constraints on form–function relationships) appears to "trump" evolutionary legacy. The first conclusion rests on the results of computer simulations which show that plant organs are multifunctional; many of the functions they perform have opposing engineering "design specifications" that must be reconciled (optimized) simultaneously; a limited number of morphological variants capable of optimizing these functions exists; and many of these "optima" reappear at the closure of "adaptive walks" through randomly shifted selection regimes.

The second conclusion rests on statistical and phyletic inferences showing that total body mass partitioning patterns among leaves, stems, and roots (or their analogs) are convergent among phyletically very different lineages. This convergence cannot be the result of "developmental constraints" *sensu stricto*, because the developmental schema underwriting multicellularity evolved independently in each lineage in which this body plan evolved. Therefore, convergence in the biomass partitioning patterns of multicellular plants must be the result of adaptive evolution in terms of functionally equivalent body parts.

However, it would extremely premature to suggest that these analyses confirm the supremacy of function over form. The use of optimization models in the study of functional morphology presupposes that an organism's form is constrained more by its environment (Darwin's "conditions of existence') than by its intrinsic organization (Cuvier's "unity of type"). A major criticism leveled against this approach is that optimization processes generally work only in systems that have reached their "equilibrium condition." This assumption is unwarranted for evolving systems, particularly those that have "memory." Organisms are historical entities. They retain ancient character states regardless of their current derivative nature. Therefore, the "design" of an organism is often a "compromise" between current environmental demands and past history. Additionally, the computer simulations presented here are based on a number of untested or problematic assumptions – for example, long-distance spore dispersal was assumed to be mediated by wind; mechanical stability was assumed to be governed exclusively by self-loading (wind-induced bending moments were entirely neglected).

Likewise, the statistical concordance in the scaling of body-part biomass across phyletically very different plant lineages may reflect "developmental constraints" operating at some as yet unidentified genomic level shared across all photosynthetic eukaryotic life forms – for example, *Hox* and *MADS* box "master" transcription factors. This possibility is consistent with the concordance between the biomass partitioning patterns observed for multicellular brown algae and the unicellular green macrophyte *Caulerpa*. If true, "historical legacy" rather than "adaptive convergent evolution" accounts for the similar biomass partitioning patterns observed for algal macrophytes, charophytes, and non-vascular and vascular land plants.

What can be said with slightly more confidence is that the antithetic argument of "form versus function" will not be resolved canonically, because the dominance of one evolutionary phenomenon over another is both time- and taxon-dependent. This somewhat pessimistic view emerges from the fact that, taken in isolation, evolutionary patterns do not explicate evolutionary processes. The analysis of patterns of organismal form, function, and their temporal or geographic distribution based on patterns of phylogenetic relationships rests on three principles: (1) organisms are ordered into nested sets defined by uniquely derived character states; (2) these sets can be used to deduce genealogical relationships among lineages; and (3) the pattern of structural relationships specified by accepting any one of a number of equally plausible phylogenetic hypotheses can be used to infer form–function and temporal or geographic patterns. However, the explicit testing of theories about the evolutionary processes has three requirements: (1) an unequivocally correct phylogenetic hypothesis (i.e. a unique and valid phylogenetic hypothesis); (2) a general theoretical framework for the relationships among form, function, and environment; but (3) one that can be fine-tuned and adapted to deal with the unique morphological attributes of different groups of monophyletic taxa. Because the evolutionary pattern of each clade is undoubtedly the result of one or more unique historical events, it is only by means of determining the relative frequency with which "historical contingency" or "adaptive optimization" dominates the general picture of life's history that we can hope to resolve the debate.

REFERENCES

Anderson, R.A. (2004). Biology and systematics of the heterokont and haptophyte algae. *American Journal of Botany* 91, 1508–22.

Andrews, Henry N. Jr., Kasper, A.E., Forbes, W.H., Gensel, P.G., and Chaloner, W.G. (1977). Early Devonian flora of the Trout Valley Formation of northern Maine. *Review of Palaeobotany and Palynology* 23, 255–85.

Bierhorst, D.W. (1971). *Morphology of Vascular Plants*. New York: Macmillan.

Bock, W.J. (1980). The definition and recognition of biological adaptation. *American Zoologist* 20, 217–27.

Bold, H.C. (1967). *Morphology of Plants*. New York: Harper & Row.

Cracraft, J. (1981). The use of functional and adaptive criteria in phylogenetic systematics. *American Zoologist* 21, 21–36.

Denny, M.W., Gaylord, B.P., and Cowen, E. (1997). Flow and flexibility II. The roles of size and shape in determining wave forces on the bull kelp *Nereocystis luetkeana*. *The Journal of Experimental Biology* 200, 3165–83.

Edwards, D. (1970). Fertile Rhyniophytina from the Lower Devonian of Britain. *Palaeontology* 13, 451–61.

Edwards, D.S. (1980). Evidence for the sporophytic status of the Lower Devonian plant *Rhynia gwynne-vaughanii*. *Review of Palaeobotany and Palynology* 29, 177–88.

Eggert, D.A. (1974). The sporangium of *Horneophyton lignieri* (Rhyniophytina). *American Journal of Botany* 61, 405–13.

Eigen, M. and Winkler, R. (1983). *Laws of the Game*. New York: Harper & Row.

Enquist, B.J. and Niklas, K.J. (2001). Invariant scaling relations across tree-dominated communities. *Nature* 410, 655–60.

(2002). Global allocation rules for patterns of biomass partitioning across seed plants. *Science* 295, 1517–20.

Gans, C. (1974). *Biomechanics: An Approach to Vertebrate Biology*. Philadelphia: J.B. Lippincott.

Gifford, E.M. and Foster, A.S. (1988). *Morphology and Evolution of Vascular Plants*. New York: Freeman & Co.

Gould, S.J. (1970). Evolutionary paleontology and the science of form. *Earth-Science Review* 6, 77–110.

(1980). Is a new and general theory of evolution emerging? *Paleobiology* 6, 119–30.

Gould, S.J. and Lewontin, R.C. (1979). The spandrels of San Marco and the Panglossian paradigm. *Proceedings of the Royal Society of London*, B, 205, 581–98.

Graham, L.E. and Wilcox, L.W. (2000). *Algae*. Englewood Cliffs, NJ: Prentice Hall.

Grant, R.E. (1972). The lophophore and feeding mechanism of the productidina (Brachiopoda). *Journal of Paleontology* 46, 213–49.

Harvey, P.H. and Pagel, M.D. (1991). *The Comparative Method in Evolutionary Biology*. Oxford University Press.

Holbrook, N.M., Denny, M.W., and Koehl, M.A.R. (1991). Intertidal "trees": consequences of aggregation on the mechanical and photosynthetic properties of sea-palms *Postelsia palmaeformis* Ruprecht. *Journal of Experimental Marine Biology and Ecology* 146, 39–67.
Kaplan, D.R. (1992). The relationship of cells to organisms in plants: problem and implications of an organismal perspective. *International Journal of Plant Sciences* 153, S28–S37.
Kaplan, D.R. and Hagemann, W. (1991). The relationship of cell and organism in vascular plants. Are cells the building blocks of plant form? *BioScience* 41, 693–703.
Keeling, P.J. (2004). Diversity and evolutionary history of plastids and their hosts. *American Journal of Botany* 91, 1481–93.
Koehl, M.A.R. and Wainwright, S.A. (1977). Mechanical adaptations of a Giant Kelp. *Limnology and Oceanography* 22, 1067–71.
Niklas, K.J. (1992). *Plant Biomechanics: An Engineering Approach to Plant Form and Function*. University of Chicago Press.
(1994). *Plant Allometry: The Scaling of Form and Process*. University of Chicago Press.
(1997a). Adaptive walks through fitness landscapes for early vascular land plants. *American Journal of Botany* 84, 16–25.
(1997b). Effects of hypothetical developmental barriers and abrupt environmental changes on adaptive walks in a computer-generated domain for early vascular land plants. *Paleobiology* 23, 63–76.
(1997c). *The Evolutionary Biology of Plants*. University of Chicago Press.
(2004a). Plant allometry: is there a grand unifying theory? *Biological Reviews* 79, 871–89.
(2004b). The cell walls that bind the tree of life. *BioScience* 54, 831–42.
(2005). Modelling below- and above-ground biomass for non-woody and woody plants. *Annals of Botany* 95, 315–21.
Niklas, K.J. and Enquist, B.J. (2001). Invariant scaling relationships for interspecific plant biomass production rates and body size. *Proceedings of the National Academy of Sciences USA* 98, 2922–7.
(2002). On the vegetative biomass partitioning of seed plant leaves, stems, and roots. *American Naturalist* 159, 482–97.
Niklas, K.J. and Kerchner, V. (1984). Mechanical photosynthetic constraints on the evolution of plant shape. *Paleobiology* 10, 79–101.
Okubo, A. and Levin, S.A. (1989). A theoretical framework for data analysis of wind dispersal of seeds and pollen. *Ecology* 70, 329–38.
Rosen, R. (1967). *Optimality Principles in Biology*. London: Butterworth.
Rudwick, M.J.S. (1964). The inference of function from structure in fossils. *British Journal of Philosophy and Science* 15, 27–40.
(1968). Some analytic methods in the study of ontogeny in fossils with accretionary skeletons. *Journal of Paleontology*, Supplement 42, 35–49.
Russell, E.S. (1916). *Form and Function: A Contribution to the History of Animal Morphology*. London: John Murray.

Santelices, B. (2004). A comparison of ecological responses among aclonal (unitary), clonal, and coalescing macroalgae. *Journal of Experimental Marine Biology and Ecology* 300, 31–64.

Taylor, T.N. and Taylor, E.L. (1993). *The Biology and Evolution of Fossil Plants*. Englewood Cliffs, NJ: Prentice Hall.

Thompson, D'Arcy W. (1917). *Growth and Form*. Cambridge University Press.

Vogel, S. (1981). *Life in Moving Fluids: The Physical Biology of Flow*. Boston, MA: Willard Grant Press.

Wainwright, S.A., Biggs, W.D., Currey, J.D., and Gosline, J.M. (1976). *Mechanical Design in Organisms*. New York: John Wiley & Sons.

4

Evolution in the light of embryos: seeking the origins of novelties in ontogeny

RUDOLF A. RAFF AND ELIZABETH C. RAFF

ROLES FOR DEVELOPMENT IN EVOLUTION

In the later part of the nineteenth century, it became evident that there was a hereditary connection between the development of living organisms and their evolutionary ancestors (Amundson 2005). This connection was given mechanistic underpinnings by Müller (1869), and was elaborated and formalized by Ernst Haeckel in his recapitulation theory, in which he attempted to link the mechanisms of evolution of descendants to the development of their ancestors by means of addition of new terminal stages, combined with the shortening of life stages inherited from the ancestors. The result, according to Haeckel, was that living forms in their development literally recapitulate the adult forms of their ancestors in a condensed sequence. This was the most influential theory of the late nineteenth century, linking evolution causally to both Lamarkian heredity and ontogeny (Gould 1977; Richardson and Keuck 2002). Although important in inspiring studies in comparative embryology, which were used in seeking phylogenetic information on remote ancestors, the hold of this theory on biologists weakened by the mid 1890s. Following Roux's influential lead, embryologists became interested in the mechanisms of developmental processes per se and abandoned Haeckel's phylogenetic program that offered little insight into how embryos develop. Haeckelian recapitulation as a mechanism was subsequently shown to be unfeasible with the rediscovery of Mendelian inheritance, which uprooted terminal addition from its theoretical

We thank Michael Ruse for inviting us to take part in this Werkmeister symposium. We acknowledge the NSF and NIH for support of our research, and thank our co-workers in Bloomington and Sydney.

foundations. In the first half of the twentieth century, evolution by Haeckel's recapitulation was discredited on both theoretical (De Beer 1951; Garstang 1922; Needham 1933) and experimental grounds (Berrill 1955; Lilly 1898). Recapitulation also became irrelevant to embryologists and geneticists struggling to understand the foundations of developmental information (e.g. Sander and Schmidt-Ott 2004).

However, despite its dethronement as a mechanistic theory for developmental biology, and the understanding that any stage of development – not only the terminal stage – could evolve, the concept of recapitulation nonetheless continued to serve as an interpretive framework for metazoan phylogeneticists. This more resistant theme of Haeckelian thought rested on the application of recapitulation theory to ideas about the evolution of larval forms and animal phylum body plans, because although there were abundant morphological and fossil data, these lacked the critical information needed to unite groups of phyla. Comparative embryological data provided some impressive links connecting phyla based on shared larval forms, and could be used to support a Haeckelian position on the evolutionary recapitulation of embryonic stages. The idea that living larval forms resemble ancestral free-living planktonic metazoans has thus had a long-standing influence, and accordingly has been an issue of importance to the design and interpretation of experiments on the evolution of marine larvae (Jågersten 1972).

Fig. 4.1 shows part of an educational chart of phylogenetic relationships published by Heintz and Stormer (1937). Even at this relatively late date, their diagram directly depicts Haeckelian ideas. Metazoans are shown to have originated from a unicellular organism (A) that acquired multicellularity (B). Once that was achieved, primitive ciliated planktonic animals evolved – animals whose morphology reflects forms that occur in modern developmental stages. Thus the *Blastaea* animal (C) was a hollow ball of cells, analogous to the blastula stage of modern embryos. Similarly, the more advanced *Gastraea* animal (D) had an invaginated gut, analogous to the gastrula stage of modern embryos. As depicted in the chart, the *Gastraea* gave rise to other more complicated planktonic animals that closely resemble the basal larvae of living protostomes and deuterostomes. The adult body plans of mainly benthic phyla are shown in the chart to have evolved from these larva-like creatures. For example, the two-layered (endoderm inner layer and ectoderm outer layer) *Gastraea* (D) was envisaged as the direct ancestor of the diploblastic cnidarians, such as the anemone (F). The *Gastraea*

Figure 4.1 The literalist recapitulationist view of evolution of phyla from embryo and larva-like ancestors. (Detail from the diagram, *Relationships of the Animal Kingdom* (Heintz and Stormer 1937): A. ancestral single cell organism (moneran). B. Early multicellular organism. C. *Blastaea*. D. *Gastraea*. The *Gastraea* has two germ layers and is directly ancestral to cnidarians (F) in this scheme. The gastraean gives rise to the ancestral protostome (G) and deuterostome (I) larva-like creatures with their distinct modes of coelom formation.

was also seen as the ancestor of triploblasic larva-like forms that had added a third cell layer, mesoderm. Of these, (G) was envisaged as ancestral to protostomes and (I) to deuterostomes, based on how mesoderm was generated in the larva-like ancestors. Even though this view was discredited on other grounds, as we will see below, the Haeckelian outlook has continued to have an influence on phylogenetic thought and on reconstruction of metazoan origins through the twentieth century.

ONTOGENY AND EVOLUTIONARY THEORY

Almost all of our direct information about evolutionary change in morphology has come from paleontology. Thus, most of our picture of evolutionary change is cast in terms of change in adult morphology over time, which is what the fossil record shows (e.g. the history of vertebrate skeletons illustrated in Gregory's magnificent *Evolution Emerging*, 1951). Despite some spectacular notable occurrences of developmental sequences, such as in Cambrian larval arthropods (Walossek 1993), fossil Precambrian and Cambrian marine invertebrate embryos (Donoghue *et al.* 2006; Xiao *et al.* 1998), or Mesozoic dinosaur embryos (Reisz *et al.* 2005), direct evidence for developmental histories is rare, and the history of morphological evolution is overwhelmingly a history of adult forms. Indeed, evolving lineages are often portrayed as series of adults or adult features illustrating changes in structures – for example, in the origin of the tetrapod limb (Shubin *et al.* 2006). This is clearly a useful shorthand. However, it is important not to mistake the shorthand for the actual mechanism. To do so would be returning to an implicit Haeckelian model. Evolution arises ontogenetically, generation by generation, from the selective interaction of ecology and behavior with development. All stages in development evolve. As the morphology of an individual arises in each generation through its ontogeny, the link between macroevolution (evolution above the species level) and microevolution (evolution in populations within a species) takes place in the evolution of developmental processes in generational steps. Development of the individual is as crucial as the generational events of genetic recombination and mutation in linking microevolutionary processes with macroevolutionary results.

We can think of the connection of development with evolution in two distinct ways. The first lies in asking how developmental processes

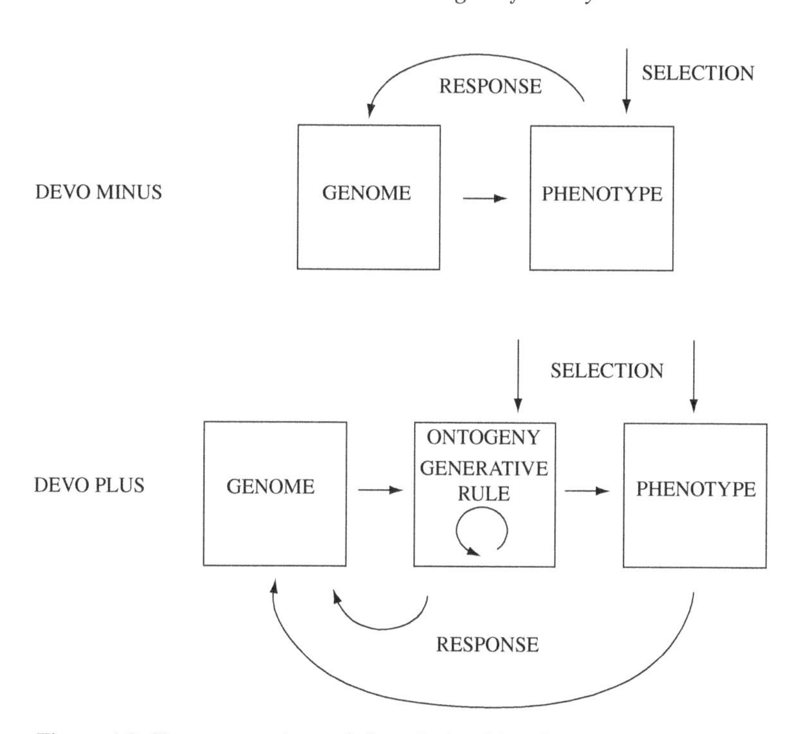

Figure 4.2 Two conceptions of the relationship of genotype to phenotype and selection. In the "devo minus" scheme, variation arises at random in the genome and is sorted as a result of selection operating on the adult phenotype. Nothing is said about how the phenotype arises or how genomic variation affects it. In the devo plus scheme, ontogeny and its generative rules are inserted between the genotype and the phenotype. Both the adult phenotype and development are subject to selection. Furthermore, it is ontogeny that converts genomic information into a phenotype. Randomly arising genomic variation is converted into non-random form by the rules operating in the existing ontogeny, a process often referred to as "developmental constraint."

evolve in a lineage, and how these changes relate to evolutionary changes in such features as body segmentation or appendage morphology. The second link lies in a consideration that developmental mechanisms per se constrain variation in non-random ways. This is not to say that mutation or other genomic changes might not occur at random, but that the mapping of phenotype on genotype is not like a number painting. Developmental processes have evolutionary histories and thus interconnected "constraints" that will determine how random gene changes can be expressed in the phenotype in non-random ways. In addition, embryos and particularly larvae are organisms themselves, and they and their ontogenies are as much targets of constraint and selection as is the adult phenotype. As we develop a more complete view of evolutionary processes, it will have to encompass not only the genotype and the phenotype, but also an insertion of developmental rules into evolutionary theory (fig. 4.2). This is a significant addition to our thinking about evolutionary mechanisms, because selection can operate at any stage of a life history, and it will operate on variation constrained by the rules of ontogeny. Ontogeny thus becomes a third "black box" in microevolution. Aspects of this issue and the evolvability of development are now being more broadly discussed (Kirschner and Gerhart 1998; Raff 1996; Wagner 1996 and papers in Hallgrimsson and Hall 2005).

MODES OF DEVELOPMENT

Historically, evolutionary studies of morphology have largely focused on features of the adult in paleontology, and in the study of function and selection on adult features of living organisms. In addition, a number of successful studies in evolutionary developmental biology (Evo Devo) have dealt with developmental mechanisms that underlie evolution of major features of adult development, such as loss of limbs by snakes (Cohn and Tickle 1999), formation of the turtle shell (Gilbert *et al.* 2001), and formation of the membrane between the fingers making up the support struts of the bat wing (Chen *et al.* 2005). For most phyla, which are largely marine invertebrates, larvae are as hugely varied as the adults and potentially provide other kinds of insights into the evolution of development.

There are two basic kinds of development recognized in terms of similarity of the body plan of larval stages to that of the adult. Further, these are broadly linked to two distinct ways in which developing individuals make the transformation to the adult condition. In taxa where larval stages possess the same body plan as the adults, as in holometabolous insects such as butterflies, the larvae are referred to as "secondary larvae." The majority of marine phyla have larvae with body plans that are distinct from the adult body plans. These are referred to as "primary larvae" (Jågersten 1972). The terminology has historical roots that give it an unfortunate coloring in terms of what is ancestral and what is derived, but the terms are well established in the literature.

Primary larvae represent developmentally and morphologically distinct phases of complex life histories. They have a fundamentally different body plan than the adult, and a complete metamorphic transition from larva to adult. For example, echinoderm larvae are bilaterally symmetric forms that swim and capture food in the plankton, whereas their adults are pentameral benthic animals that are largely predators or grazers on the sea floor. There is no hint of the pentameral adult body plan in the larva except in the rudiment of the adult, which grows within the larva. This kind of development is called "indirect development." Echinoderms are not the only indirect developers: other examples of indirect-developing clades include marine mollusks, bryozoans, annelids, hemichordates, and indeed the majority of the thirty-five or so animal phyla. Animal life originated and underwent its diversification into major adult and larval body plans in the sea, and most of these indirect-developing organisms, with their distinct larval forms, are marine. Indirect developers generally have small eggs and produce planktonic feeding larvae.

The second mode of ontogeny is called "direct development," in which there is no fundamentally distinct larval body plan. Humans and the nematode *C. elegans* are direct developers, with only a few features that could be regarded as distinctly non-adult. There is no distinct break in body-plan morphology during growth, although changes in body proportion and sexual maturation occur. The picture is somewhat complicated because some direct developing clades have evolved larval features in the pre-adult stages of some of their derived sub-clades. These "secondary larvae" can be understood by a familiar example. Vertebrates are primitively direct developers. However, the frogs, a derived clade, have evolved secondary tadpoles, larvae analogous to those of holometabolous insects. Frogs are direct developers in that their larval form has the canonical vertebrate body plan, even though the tadpole, a secondary larva, differs in significant ways from the adult – that is, by possession of a swimming tail, modified gut, and external gills, by lack of limbs, and by a metamorphic process that replaces larval with adult features at a particular time.

To confuse matters a bit, a number of marine invertebrate taxa with primitively indirect-developing primary larva have later evolved larvae that have reduced most of their larval traits and develop directly into the adult body plan. Direct developers often have large eggs and no need of larval feeding. There is frequently some conflation in discussions between developmental mode and trophic strategy, which do not always cleanly segregate in larval evolution (McEdward and Janies 1997). For example, there are species that produce a larva with feeding structures, but with sufficient food reserves that they do not need to feed. Such forms may themselves be evolving into direct developers.

Although we generally consider the adult to be the definitive form of a species, the study of the evolution of larval morphology provides an experimentally simpler system for understanding how morphological novelties arise in ontogeny, albeit in a different life history phase from the adult. In real terms, all phases of a life cycle are important, and in marine invertebrates, most dispersal in the sea involves larvae, not adults. More importantly, most Evo Devo studies of developing organisms, even when conducted on earlier developmental stages, have dealt with arthropods and vertebrates. Although the "secondary" larvae in these groups represent anatomically and ecologically distinct entities, these phyla are primitively direct developers; that is, their larval forms have the same body plans as the adults. The phyla that make "primary" larvae have genomes that contain the information to construct a second body plan beside the adult one.

HOW DID LARVAE EVOLVE?

Although the mechanistic props of Haeckelian recapitulation were knocked out early in the twentieth century, the last half of the twentieth century nevertheless saw interesting elaborations of the Haeckelian view of larval origins and their applications to modern research programs. The phylogenetic history in the chart in fig. 4.1 was constructed by giving a recapitulationist interpretation to the ontogenies of living forms, in which developmental stages of extant organisms were seen as being derived from actual primitive adult animals. In a more recent application of this view, Jågersten (1972), in *Evolution of the Metazoan Life Cycle*, summarized his *Bilaterogastrea* theory in this way: "According to this theory the two phases of the life cycle arose when the adult of the primeval ancestor of the metazoans, viz, the holopelagic, radially symmetrical *Blastaea*, descended to life on the bottom (and became bilateral), while *its juvenile stage remained in the pelagic zone.*" He goes on to suggest that initially differences between the two phases was slight. Little metamorphosis was necessary in these early hypothetical animals. There was a gradual evolutionary divergence of a benthic adult from the pelagic larva as evolution proceeded. Primary larvae are thus considered basal to metazoan phyla. Nielsen and Nørrevang (1985) and Nielsen (1995) have suggested that a pelagic *Gastraea* animal gave rise to a pelagic *Trochaea* animal (that is, resembling a particular type of feeding larva), ancestral to subsequent great clades of protostome and deuterostome phyla.

The most recent and interesting application of the Haeckelian perspective lies in its combination with inferences about gene regulatory systems (Davidson *et al.* 1995; Peterson *et al.* 1997). This approach posits primitive Precambrian metazoans as larval-sized pelagic organisms. The gene regulatory systems of these animals are hypothesized to have resembled those found in living marine larvae. The ancestral metazoans are thus envisioned as larva-like both in form and genetic regulation. The origin of bilaterian adults is suggested to have occurred through the invention of imaginal "set-aside" cells distinct from the majority of differentiated larval cells. It is from these set-aside cells that the adult rudiment arises, and emerges at metamorphosis. Gene regulation in the new adults was hypothesized to be based on a gene regulatory system similar to those of living adult bilaterians, including the use of Hox genes to pattern the anterior-posterior body axis. This is unlike the case of primary larvae, in which *Hox* genes play little role. There are evident difficulties for this interlinked suite of hypotheses, notably the vast number of convergent events required; accounting for the massive molecular convergences in *Hox* use in independently evolved descendent clades with benthic body plans; identifying a selective role for set-aside cells before a new bilateral and benthic adult stage has evolved; selection for the novel developmental elements prior to need; and incongruencies with phylogenetic inferences.

To infer phylogenies, the Haeckelian approach used larval forms and the hypothesis that these represented developmental repetitions of ancestral forms. Newer methods independent of embryos, including DNA sequence-based methods, produce phylogenies that invert the "embryos first" conclusion. Clades with highly distinct "primary" larval forms are likely not basal in metazoans (Jenner 2000; Rouse 2000; Sly *et al.* 2003). If we ask where indirect development maps on the metazoan phylogenetic tree, it appears that the most basal clades are primitively direct developers, and that indirect development via "primary" larvae arose later in evolution. Phylogenetic considerations indicate that indirect development evolved independently in several lineages: in the protostomes primarily in the lophotrochozoan clade (animals with a trochophore larva, like mollusks and annelids, plus phyla that feed with a lophophore, like brachiopods and bryozoans); and in the deuterostomes the echinoderm plus hemichordate clade (Jenner 2000; Sly *et al.* 2003).

The late Precambrian and early Cambrian fossil record implies that adult ancestral bilaterian metazoans were small – although by no means necessarily microscopic (Donoghue *et al.* 2006). In consequence, they were direct developers, as small animals cannot produce the large number of eggs needed for planktotrophic indirect development. It is not known whether these early bilaterally symmetric animals (bilaterians) were pelagic or benthic, but worm-like small benthic bilaterians are common. The most primitive relative of echinoderms and hemichordates is *Xenoturbella*, exactly such a little worm, which in addition broods its embryos within the parent's body (Bourlat *et al.* 2003; Israelsson and Budd 2005). Thus, we can conclude that the ancestor shared by *Xenoturbella* and vertebrates lacked a "primary" larva. The subsequently evolved common ancestor of the living echinoderm plus hemichordate clade had feeding planktonic larvae.

Primary larvae can only have originated when early bilaterians evolved body sizes sufficiently large to allow them to produce enough gametes to support planktotrophic development, because vast numbers of planktonic larvae are eaten before metamorphosis. Primary larvae hypothetically evolved through interpolation of features such as cilia (the basis of both locomotion and feeding) or a gut into early stages of direct development of a benthic bilaterian ancestor, to yield features associated with feeding larvae (Sly *et al.* 2003; Valentine and Collins 2000; Wolpert 1999). The genetic basis for interpolation of new features is the co-option of genes already used in the adult into an earlier stage of development by change in the regulatory regions of these genes. The genes involved would likely evolve cis-regulatory sites that would permit transcription in early development, in addition to continued adult expression. This is analogous to the recruitment of various ordinary metabolic enzymes as eye lens proteins "crystallins" in various clades. These recruited genes gain transcriptional regulatory sites that allow the proteins they encode to be expressed at high levels in a novel venue for a novel function as close-packed globular proteins, which provide a clear and stable medium that transmits light through the lens (Piatigorski 2003).

The paleontological record shows that large-bodied bilaterians appear as fossils in the early Cambrian, ca 545 mya, rapidly evolving into the fantastic lobopods, trilobites, annelids, priapulids, anomalocarids, primitive echinoderms, basal vertebrates and other forms so well displayed in the Burgess Shale and Chengjiang faunas of the mid-Cambrian. Other fossil evidence (trace fossils of bilaterian animals creeping on the sea floor and probable fossil embryos) suggests that late Precambrian bilaterians existed by 580 mya. The adults were likely small, lacked any skeleton, and were unlikely to fossilize. The earliest fossil skeletal elements from that time suggest small body sizes. The sizes of fossil embryos are consistent with a dominance of direct developers prior to the Cambrian (Peterson 2005, 2007; Raff *et al.* 2006; Dunn et al. 2007; Nützel *et al.* 2006).

Although paleontology can reveal little of the mechanisms underlying the events of larval origins, the hypothesis that indirect-developing feeding larvae arose through gene co-option and intercalation into early development is testable experimentally using living larval forms. To document co-option, one needs to identify in living lineages genes that are candidates for co-option into larval functions. We have found many examples of such gene co-option candidates in sea urchins. These can be studied to define the kinds of molecular and genetic mechanisms involved in co-option events during larval evolution, and by extrapolation to inform us about gene regulatory events in the Cambrian origins of primary larvae.

EXPERIMENTAL STUDY OF RAPID LARVAL EVOLUTION

In this section we address origins of larval forms as actively evolving entities that provide an accessible experimental system in which to dissect the molecular and cellular mechanisms that underlie evolutionary change in body form and function. We will approach larval evolution from two perspectives. First is the hypothesis that so-called primary larvae do not represent primitive ancestral forms, but rather arose secondarily after the origin of the suite of extant body plans, and continue to undergo sometimes rapid evolutionary change. As outlined above, the origin of primary larvae occurred in the late Cambrian to early Ordovician, long after the Precambrian radiation that gave rise to the basic body plans of the basal representatives of the earth's extant animal phyla (Budd and Jensen 2000; Dunn *et al.* 2007; Nützel *et al.* 2006). These primary larvae have evolved important features after the initial evolution of feeding larvae, providing a window onto the evolution of primary larvae. Second, we will describe our studies of a recently evolved larval form in the sea urchin *Heliocidaris erythrogramma*, which has geologically recently evolved a form of secondary direct developing larva from a form with a pluteus larva. These studies have allowed us to determine some of the mechanistic processes that have produced dramatic and evolutionary rapid changes in early development and resulting larval morphology.

The fossil record of echinoids, sea urchins, seems to start in the late Ordovician ca 450 mya. Echinoderms are 80 myr older in the fossil record, and their ancestors had already acquired feeding "primary" larvae by the time sea urchins evolved. Sea urchins have a characteristic larva, the pluteus (characteristic of sea urchins with indirect-developing feeding larvae), which has derived features not present in more basal echinoderm feeding larvae or the hemichordate plus echinoderm ancestral larva (Harada *et al.* 2002; Nakajima *et al.* 2004; Primus 2005; Smith 1997). Phylogenetic patterns indicate that the pluteus was present 250 mya, by the end of the Paleozoic. Pluteus larvae possess long ciliated arms, used in both locomotion and feeding. Each arm is supported by a calcite skeletal rod. These arms constitute a derived feature of the pluteus, and are not present in the basal echinoderm feeding larva (fig. 4.3A). These observations are also emblematic of the fact that larvae evolve, and that Haeckelian recapitulation does not offer a useful model.

We have demonstrated that a set of genes has been co-opted for use in the developing larval arm in the echinoid lineage. These genes include members of the tetraspanin, advillin, and carbonic anhydrase families (Love *et al.* 2007). These genes are expressed in adult sea urchins, but are specifically expressed in the ectoderm (tetraspanin) or mesenchyme (advillin and carbonic anhydrase) of the growing arm tips, here shown in the Australian sea urchin, *Heliocidaris tuberculata* (fig. 4.4). A number of conclusions have emerged. The arm is a novel larval organ, and arms arose after the divergence of echinoids from other echinoderms. As theses genes existed prior to the evolution of the pluteus, the genes were co-opted to be expressed in specific larval cells, and to be involved in formation of the novel larval structure. The co-option event occurred in the echinoid clade. Thus, gene convergence in the evolution of new larval features tells us not only that co-option of genes for new larval features continued beyond the events leading to primary larvae, but also that such genes can be identified and the events that are involved in gene co-option into producing a novel feature can be studied.

Larval evolution did not cease hundreds of millions of years ago in the Paleozoic era. It continued into geologically recent times (an excellent review of all aspects of larval evolution is presented by Wray 1995). Thus, evolution of non-feeding sea urchin larvae has occurred in living clades of sea urchins within the past four to seven million years

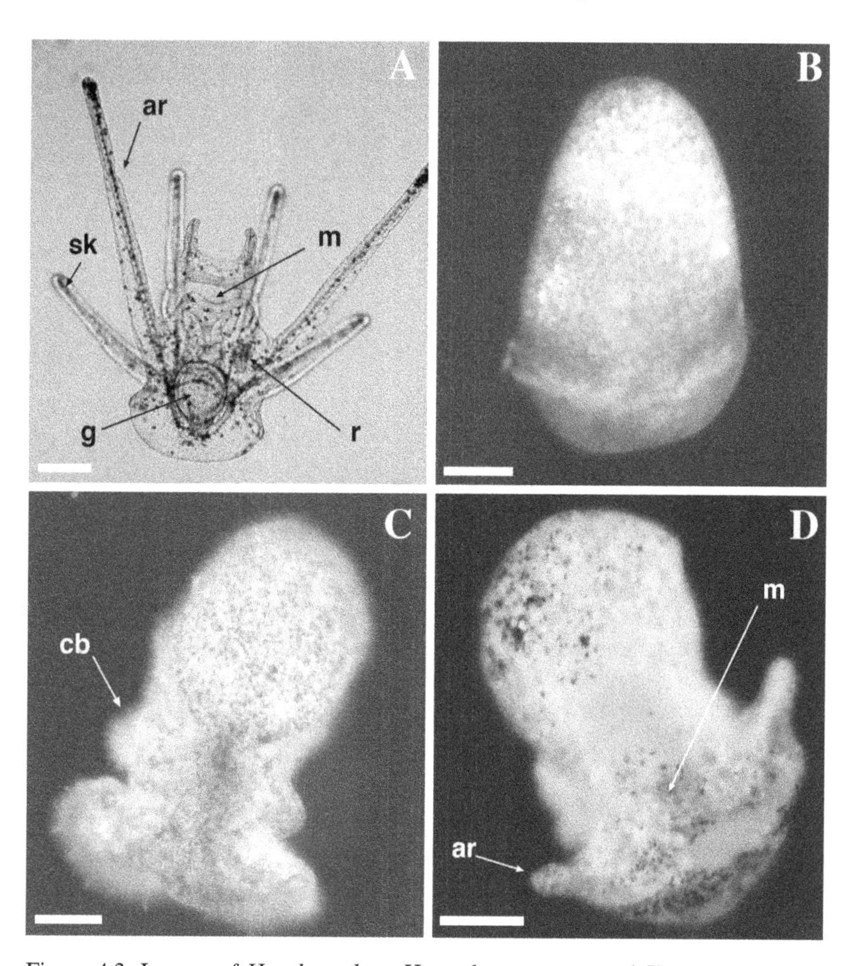

Figure 4.3 Larvae of *H. tuberculata*, *H. erythrogramma*, and *H.e. x H.t.* hybrids generated from *H. erythrogramma* eggs fertilized with *H. tuberculata* sperm. A. Ventral view of a seventeen-day-old eight-armed *H. tuberculata* pluteus, showing the larval arms (ar), mouth (m), and opening of the gut (g). The mass on the left side of the pluteus is the developing adult rudiment (r). The internal calcareous skeleton of the arms is visible through the transparent larval ectoderm (sk). Metamorphosis to generate the juvenile adult occurs at approximately six weeks. B. Left ventral view of a three-day *H. erythrogramma* larva. The direct-developing larva does not have arms, a functioning gut, or a complete ciliary band. Metamorphosis occurs at three–four days. C. Dorsal view of a three-day *H.e. x H.t.* hybrid larva, showing the deeply lobed body encircled by a complete ciliary band (cb), homologous to the ciliary band that encircles the arms of the pluteus. D. Ventral (oral) view of another three-day hybrid, showing the stomodeal invagination that becomes the mouth (m), and the developing pluteus-like skeletonized larval arms (ar). Metamorphosis of the hybrids occurs at seven days. Scale bars, 100 μm. After Raff *et al.* (1999).

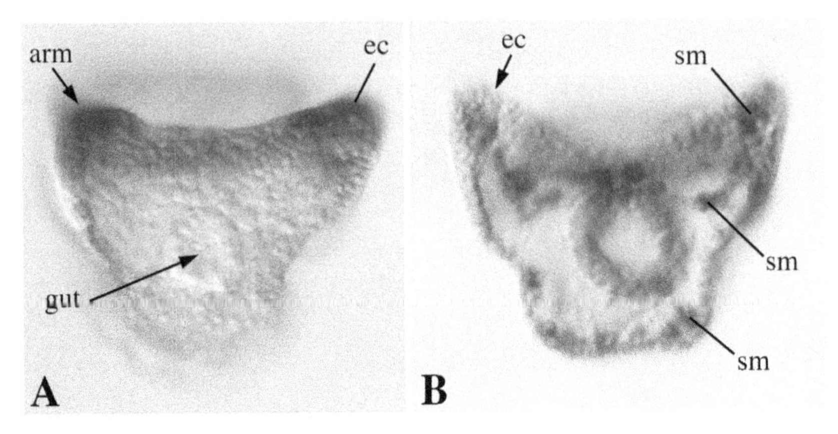

Figure 4.4 Gene expression in the growing arm tips of the *H. tuberculata* pluteus larva. A. View of an early pluteus. The *in situ* hybridization signal for the gene *tetraspanin* is strongest in the ectoderm of the arm tips. B. View of a similar larva showing expression of the *advillin* gene, which is expressed in the skeletal mesenchyme that lies beneath the arm tip ectoderm and in other sites of skeletal growth. Arrows indicate locations. Lines show sites of *in situ* hybridization labeling. Abbreviations: arm = arm tip; ec = ectoderm; gut = optical section through the gut; sm = skeletal mesenchyme.

(Jeffery and Emlet 2003; Jeffery *et al.* 2003; Sly *et al.* 2003). We can make a prediction using these recently modified larval forms: if arms were subsequently lost during further larval evolution, the coordinate expression of these genes should be lost as well. This situation exists with a sister species, *H. erythrogramma* (fig. 4.3B). This species diverged from *H. tuberculata* about four mya (Zigler *et al.* 2003). In this case, the lineage of *H. erythrogramma* has lost a number of pluteus features present in its shared ancestor with *H. tuberculata*, such as a gut and arms. *H. erythrogramma* is a non-feeding direct developer. All of the genes associated with arm development are expressed in *H. erythrogramma*, but coordinate regulation of expression of arm-expressed genes has been lost with arm loss. Thus, early developmental features are readily "evolvable," as are the gene regulatory networks that underlie particular larval features. *H. erythrogramma* has been immensely useful as a non-"model system" model for the evolution of larval development. This looks like a loss of complexity, but in fact, the apparently simplified morphology of this larva conceals complex underlying process changes and formation of a new morphological entity, not simple degeneration. Changes in developmental gene regulatory systems have been dramatic, with gene regulatory co-option a continuing process in larval evolution.

CONVERGENT EVOLUTION OF DIRECT-DEVELOPING LARVAE IS COMMON IN SEA URCHINS

Convergence is the independent evolution of similar features. This can be through similar genetic means – that is, parallelism – or through unrelated genetic and developmental mechanisms – that is, classic convergence. Current evolutionary studies, notably those revolving around the resolution of phylogenetic problems, center on identifying homologies and understanding their modifications in diverging lineages. However, despite the branching pattern of phylogenetic trees, convergence is prevalent in evolution, and results in some of the remarkable adaptations seen in nature – for example, the swimming apparatus of whales as contrasted with that of fish, the wings of birds and bats, and the features of direct-developing larvae among sea urchin lineages. The evolution of non-feeding direct-developing larvae has occurred independently in a number of sea urchin orders (Wray 1996). Analogous evolution of direct developers from indirect developers has occurred in many clades among echinoderms and other phyla, marking a prevalent kind of convergence in marine invertebrates. Among sea urchins, each ordinal lineage primitively has a pluteus larva. Direct developers have arisen independently in several lineages. Despite the evolutionary distance between the lineages, remarkably similar direct-developing forms to that of *H. erythrogramma* have evolved in each. Our studies provide a suggestion that these morphological convergences might well lie in similar independent modifications of long-conserved genic mechanisms. This view is consistent with the demonstration by Hinman *et al.* (2003) that major links within developmental gene regulatory networks are conserved between sea urchins and starfish embryos, over the nearly 500 myr since their divergence.

A HYBRID STRATEGY TO GAIN INSIGHTS INTO REGULATORY CHANGES IN EVOLUTION

Simple experimental techniques can be highly revealing. We have made use of a century-old classic technique in sea urchin embryology for a new purpose. That is, with many pairs of sea urchin species one can produce cross-species hybrids. In the laboratory we can fertilize eggs of one species by sperm of another, even in cases where this has little chance of occurring in nature. We gave hybridization a new twist by using two parental species of different developmental modes. We asked if hybrids of species of different ontogenies can be made and what features will appear in the resulting hybrids. The important cross for this effort was that in which eggs of the direct developer *H. erythrogramma* were fertilized by sperm of the indirect-developing *H. tuberculata* (H.e. x H.t.). The phenotype of hybrid embryos was a surprise in being coherent in pattern. It was also distinct from that of either parental species (fig. 4.3C, D) (Raff *et al.* 1999). Most surprisingly, these larvae, generated by a novel ontogeny, metamorphosed into juvenile adult sea urchins.

We drew a number of conclusions from *H.e. x H.t.* cross-species hybrids. The first was that the ontogeny of the hybrids was not only novel, it was also robust. The perturbations in gene expression resulting from the combining of two genomes that in the parental species produce distinct ontogenies should have been highly disruptive. Yet the hybrid genotypes gave rise in combination to a coherent ontogeny. We know that both genomes are expressed in the hybrids, and clearly their integrated regulation takes place (Nielsen *et al.* 2000; Wilson *et al.* 2005a). The effect was not wholly genetic. The cross in the opposite direction (*H.t. x H.e.*) develops to gastrulation and then fails to make proper larval body axes and dies. Thus, maternal effects arising from the *H. erythrogramma* egg supply organizing information needed for successful development. The course of hybrid ontogeny was important in another respect. The hybrid ontogeny shows the role of dominant regulatory factors derived from the *H. tuberculata* paternal parent. The hybrids first express *H. erythrogramma* cleavage patterns and morphology, but then begin to shift toward a pluteus-like morphology, including restoration of the oral ectoderm, larval arms and skeleton, and apparently functional gut.

We were also able to make cross-species hybrids between *H. erythrogramma* and phylogenetically distant indirect-developing species, as well as hybrids with a distant, independently evolved direct developer. We used hybrids of these types to ask questions about rates of evolution and convergent evolution (Raff *et al.* 2003). A major issue that we addressed in this way was to determine how disparate are rates of evolution of larval developmental regulatory systems. The paleontological data that morphological evolution can be punctuational imply that ontogenies evolve in a punctuational way during such events. However, no direct demonstration of punctuational evolution of a developmental regulatory system had been made. The deep conservation of pluteus morphology suggests a similar stasis of the regulatory pathways that govern its development. In contrast, *H. erythrogramma* has diverged rapidly and considerably from the canonical pluteus in the four myr since its split from *H. tuberculata*, which retains a typical pluteus, and thus morphological stasis. We devised a test for the hypothesis that developmental regulatory evolution has been fast in *H. erythrogramma* but slow in species developing via a pluteus. This was done using cross-species comparisons of different phylogenetic distances. The *H.e. x H.t.* cross is between species separated by four myr. The resulting phenotype was compared to a second cross of *H.e. x P.m.*, where *P.m.* is *Pseudoboletia maculata*, a forty myr diverged member of another family – a large phylogenetic distance. *P. maculata* also has a typical pluteus. Our hypothesis was that as most of the evolutionary change had accumulated in the *H. erythrogramma* lineage, the *H.e. x P.m.* cross should produce hybrids very similar to those of the *H.e. x H.t.* cross. This was indeed the case for these long phylogenetic distance hybrids, supporting the hypothesis (Raff *et al.* 2003).

As only *H. erythrogramma* has been studied at the molecular level, it is not possible to directly show convergence in gene regulation with other independently evolved direct developers. We were able to test the possibility of convergence by cross-species hybridization between *Heliocidaris* species and the seventy myr removed direct developer *Holopneustes purpurescens* (Raff *et al.* 2003). We found that hybrids of *Holopneustes purpurescens* eggs fertilized by *H. tuberculata* sperm cleaved but failed in morphogenesis, indicating that genic mechanisms have diverged too much in that time to form a viable hybrid. However, hybrids between *H. erythrogramma* and *Holopneustes purpurescens* developed through metamorphosis to a juvenile sea urchin. Development was entirely via a direct-developing larva. The data strongly suggest that developmental mechanisms are highly convergent, but only characterization of developmental mechanisms in *Holopneustes* can fully test that suggestion.

MECHANISTIC DIVERSITY IN THE EVOLUTION OF NOVEL LARVAL FEATURES

Historically, there has been a desire for a unifying concept of evolutionary change in development. For most of the twentieth century, that mechanism was provided by the concept of heterochrony, in which the timing of one process in development is evolutionarily displaced from its ancestral relationship to the timing of another process (Gould 1977). Developmental genetics has made gene-based mechanisms accessible to experimental study, and has added a great deal of detailed knowledge of developmental evolution, particularly at macro scales. However, there has been remarkably little study of heterochrony at a genic level, with the notable exception of aspects of nematode temporal patterning (reviewed by Slack and Ruvkun 1997).

Another approach has been to look for major regulatory genes that play a role in morphological changes at small phylogenetic distances, often on the scale of a few thousand years in the splitting of populations or of domestication. Other studies extend to longer timescales associated with speciation (Leroi 2000; Orr 2005). Current theory suggests that large-effect mutations can be favorable and are substituted early in adaptive evolution, with smaller-effect mutations later. These predictions are borne out in experimental studies. In the past 10,000 years, stickleback fish in lakes have diversified in morphology from their salt-water ancestors, although they have not diverged as species. The genetic technique of quantitative trait locus mapping (QTL mapping), where regions of chromosomes that carry genes involved in the phenotypic trait of interest are identified is often used to find large-effect genes. In stickleback fish, QTL mapping suggests that a small number of loci account for most phenotypic variation, with one major QTL (encoding the transcription factor Pitx1) and at least four minor ones involved in pelvic reduction in lake forms (Peichel 2005; Shapiro *et al.* 2004). Over a longer interval, two closely related species of monkeyflowers, *Mimulus lewisii* and *M. cardinalis*, have diverged in color, shape, and other pollinator-related characters. A small number of QTLs show large effects, and a number show smaller effects (Bradshaw *et al.* 1998). One gene controlling color has been shown to largely control pollinator choice (Bradshaw and Schemske 2003).

The disparity in larval development between *H. tuberculata* and *H. erythrogramma* has evolved since their divergence about four mya, and is comparable in scale to other studies between closely related forms. This has allowed us to tease out evolutionary changes in major developmental processes in which *H. erythrogramma* is transformed. It has been possible to do that because of the enormous body of work on the developmental biology of indirect-developing sea urchins. This is especially notable with regard to studies of the molecular bases of the processes that give rise to the three larval body axes and differentiation of cells along these axes.

DEVELOPMENTAL-GENETIC APPROACHES TO ECHINODERM LARVAL EVOLUTION

Studies of Evo Devo have until recently largely focused on understanding gene regulatory changes at the macroevolutionary scale, involving distant clades, such as in the role of *Hox* genes expression patterns in body plan reorganization (e.g. in the evolutionary remodeling of the snake body axis: Cohn and Tickle 1999). Changes in developmental-genetic features of echinoderms at macroevolutionary levels have been made. Hinman *et al.* (2003) determined that many of the gene networks that underlie larval development have been retained between starfish and sea urchins since the divergence of these two lineages. Echinoids produce a larval skeleton and starfish do not. Hinman *et al.* (2003) observed that a regulatory gene involved in skeletogenesis in the sea urchin is used differently in the starfish, responding in starfish to endomesodermal inputs that do not affect it in the sea urchin embryo. Larval network conservation as a potential universal phenomenon among echinoderm classes is counterbalanced by evolutionary changes in the deployment of major regulatory genes (Lowe and Wray 1997).

Closely related species with divergent ontogenies can be used to investigate evolutionary pathways in larval evolution. This approach is being used with several starfish and sea urchin genera (Byrne 2006; Byrne *et al.* 2003; Hart *et al.* 1997; Jeffery and Emlet 2003; Jeffrey *et al.* 2003; Love and Raff 2006; Raff 1996; Raff *et al.* 2003; Smith 1997; Wilson *et al.* 2005a, b). The study of developmental evolution in echinoids rests on comparisons to a large amount of functional developmental biology, carried out over several decades with indirect-developing echinoids.

The ontogeny of *H. erythrogramma* has been substantially modified relative to indirect-developing sea urchins. These modifications start with maternal provisioning of the hundredfold greater volume egg and changes in maternal effects on development (Villinski *et al.* 2002). Most importantly, the primary embryo axes are established maternally as well (Henry and Raff 1990). In indirect-developing sea urchins, the animal-vegetal axis is the only axis to be determined maternally, with the oral-aboral = dorsal-ventral (D-V) and subsequently left-right (L-R) axes being established in the embryo (Duboc *et al.* 2004, 2005). Thus, in *H. erythrogramma*, there has been substantial reorganization of the maternal "program," including many processes that normally occur in the embryo. These become specified during oogenesis rather than after fertilization. However, there have also been substantial changes in the larval phase of development. The cell fate map is highly modified from those of indirect-developing sea urchins (Wray and Raff 1989, 1990). Cell cleavage patterns of early-stage embryos are modified such that there is no production of micromeres by *H. erythrogramma.*

The differences in these features are of a scale commensurate with differences in properties of embryos on the phylum level. That this variation occurs within a single genus shows that such apparent evolutionary stability does not preclude rapid changes under selection. The large effects observed in *H. erythrogramma* result from regulatory changes in otherwise conserved mechanisms that execute axial differentiation. Expression of the A-V axis in *H. erythrogramma* uses the same Wnt-8 signal transduction pathway as demonstrated in indirect development (Angerer and Angerer 2003; Kauffman and Raff 2003). In indirect developers, the other two larval axes, D-V and L-R axes, are established through the upstream action of the Nodal signaling pathway, and in part the downstream transcription factor goosecoid (gsc) (Angerer *et al.* 2001; Duboc *et al.* 2004, 2005; Flowers *et al.* 2004). The Nodal signaling system is thus used twice to generate two distinct axes. Nodal is expressed first on the ventral side and creates a signaling gradient that establishes the dorsal-ventral axis (Duboc *et al.* 2004). Nodal subsequently is expressed on the right side, but is involved in establishment of the L-R axis by repression of left-side differentiation on the right (Duboc *et al.* 2005).

We have found that in *H. erythrogramma*, the Nodal signaling pathway operates in the execution of both these axes similarly to its operation in indirect-developing sea urchin embryos, but the events are heterochronically shifted (Minsuk and Raff 2005; Smith *et al.* 2008). The appearance of the L-R axis in *H. erythrogramma* occurs much earlier in development than it does in the indirect developers: at the end of gastrulation, roughly twenty-four hours after fertilization, versus in the feeding larva about a month after fertilization (Ferkowicz and Raff 2001). This heterochronic change involves the same regulatory genes as in the axes of indirect developers, but operating in different developmental stages and in different regulatory environments (Ferkowicz and Raff 2001; Smith *et al.* 2008; Wilson *et al.* 2005a, b).

Furthermore, expression changes in gsc are related to the modifications leading to the non-feeding larval form (Wilson *et al.* 2005a, b). Manipulation of expression of gsc mRNA in *H. erythrogramma* results in an enhanced oral-aboral axis. This result indicates that an evolutionary reduction in expression of the gsc protein in presumptive oral ectoderm supplies a major genic change underlying the shift in larval morphology.

Changes in the control of the timing of axis formation is thus important in the evolution of the *H. erythrogramma* larva. The most dramatic heterochronic shift is the switch of axis determination to maternal control. This initial preformation of axial determination may allow the very fast development observed in *H. erythrogramma*. On the other hand, the loss of differentiation of an overt oral (ventral) ectodermal face in *H. erythrogramma*, including loss of the larval mouth, appears to involve the initial molecular steps in formation of the oral domain. The aborted development of oral ectoderm surrounding the larval mouth is consistent with the early loss of expression of a major transcription factor, gsc, in the oral ectodermal domain of the early larva. Following the formation of a reduced D-V axis, the formation of the larval L-R axis takes place precociously, and involves gene expression heterochronies that initiate the early second expression domain of the Nodal pathway into early larval development. Together, these events drastically speed development, nearly abolish oral side differentiation and morphogenesis, and accelerate differentiation and morphogenesis of structures along the L-R axis that begin the development of the juvenile adult rudiment. The result is that *H. erythrogramma* metamorphoses in three to four days, as opposed to about a month to six weeks required for indirect-developing species.

Finally, although novel evolutionary features are generated through heterochronic changes in the operation of conserved regulatory systems, there also are novel features as well as reductions in developmental features and gene expression. Among the novel features are the fusion of the formerly separate oral and aboral ectoderms of the pluteus into a novel "larval" ectoderm in *H. erythrogramma* (Love and Raff 2006). In contrast, a number of structures found in the pluteus, such as the arms, gut, and specialized aboral ectoderm, have been lost in *H. erythrogramma*. Genes specifically expressed in these structures are either no longer expressed in the larva, are expressed but in other patterns, or have ceased to be transcribed and as a result of mutations have "died" as genes, although they are still retained in the genome as pseudogenes

Table 4.1. *Evolutionary events in H. erythrogramma and inferred mechanisms*

Feature	"Cause"
Egg size/lipids	heterochronies, rate/duration
Maternal axis determination	heterochrony
Cleavage pattern	change MT/motor localization
Cell fate	heterochrony +?
Execution of axes	heterochrony
Early development of coelom	heterochrony
Rapid development of rudiment	heterochrony
Loss of oral field	reduced gsc expression time
Loss of larval arms	loss of gene expression regulation
Loss of larval gut	heterochrony to late expression
Novel larval ectoderm	fusion of gene expression modules
Novel expression of genes in ectoderm	co-option/gene regulation changes

(Kissinger and Raff 1998; Love and Raff unpublished data). We conclude that changes in regulation of processes affecting site, timing, and levels of gene action are all involved in rapid larval evolution, with a surprising preponderance of heterochronic changes (table 4.1).

CONCLUSIONS

The impact of Haeckelian recapitulation was of course strongest in the nineteenth century, before it was challenged on mechanistic grounds, but surprisingly, it continued to be applied through the twentieth century. The influence of Haeckelian recapitulation may well have persisted because a von Baerian form of recapitulation does occur in the development of many taxa. This kind of recapitulation is reflected in the similarities of developmental events observed among animals that possess similar body plans. Thus, mammals can be seen to recapitulate structures or events seen in the development of more basal vertebrates, and indeed a body plan with a number of shared features is present in vertebrate development. Von Baerian recapitulation reflects only conservation of some features of a shared developmental evolution, not the terminal addition of new features embodied in Haeckelian recapitulation.

For phylogeneticists, recapitulation as a form of evolutionary data remained useful because so few morphological characters united phyla, but a number of larval features did. One might or might not accept Haeckel's mechanistic ideas of evolution, but the idea that early development had to be more conserved because most evolution occurred later in development allowed for recognition of relationships that the adults did not reveal. For example, among phyla such as the segmented annelid worms and non-segmented mollusks such as snails and clams, which share few adult features, the fact that they do share a trochophore larva allowed recognition of their phylogenetic relationship. It further reinforced the idea that early stages of development evolve less than adult stages. Seeing such larval forms as ancestors allowed phylogenetic schemes to be drawn without reference to genetics, and supported a long tradition of hypothesizing about larval evolution. A few molecular developmental biologists have applied a Haeckelian perspective in building their hypotheses. For example, in a paper dating from the dawn of molecular developmental biology, Whiteley *et al.* (1970) suggested that the sequences of RNAs expressed very early in development should be more conserved than those expressed later in development. They tested this hypothesis by DNA hybridization studies in sea urchins and concluded that egg RNAs were more conserved than those of the pluteus larva.

The case made for a resemblance between marine larval gene regulatory systems and those of ancestral metazoans is especially interesting, because it does posit that ancestral metazoans actually were larva-like in morphology and in underlying developmental regulation (Davidson *et al.* 1995). This hypothesis suggested that ancestral metazoans were planktonic and that living larvae recapitulate the adults of these ancestral forms. In this scenario, all evolution to produce bilaterian phyla lies in evolution of a new terminal adult stage. This hypothesis links recapitulation to a concomitant postulate of a highly conservative system of gene regulation in marine larval forms. The long-term conservation of larval forms – for example, trochophores or dipleurulas – shows that larvae can be highly conserved. Yet larvae do evolve, in some cases very rapidly. Cases like *H. erythrogramma* show that larval gene regulatory systems can readily evolve, and can produce highly modified larval forms that do not recapitulate something ancestral. Thus, any Haeckelian recapitulation appears unlikely, with selection probably controlling both conservation and radical evolution of larvae. The study of these rapidly evolving larval forms gives us a window onto the evolution of evolutionary novelties in a simpler setting than those provided by most developmental pathways to the adult stage.

Haeckel was a crucial part of the historic and intellectual tradition of Evo Devo and its origins. Larval evolution and phylogeny has been pretty much the last bastion for recapitulation, but the emerging details of evolution in *H. erythrogramma* and other larvae show that this viewpoint is now purely of historical interest, and no longer serves as a source of insight into larval evolution.

We now have insights into a variety of mechanisms, including co-option of genes to new uses, duplication of genes and divergence in their functions, and evolution in gene regulatory regions that result in changes in place (heterotopy), timing (heterochrony), and amount of expression. There are also important integrative phenomena centered on modularity of developing organisms (Raff 1996; Wagner 1996). The roles of this new suite of evolutionary mechanisms in Evo Devo can be studied effectively by a combination of comparative biology with the tools of genomics and developmental genetics (Love and Raff 2003; Raff and Love 2004).

REFERENCES

Amundson, R. (2005). *The Changing Role of the Embryo in Evolutionary Thought: Roots of Evo-Devo*. Cambridge and New York: Cambridge University Press.

Angerer, L.M. and Angerer, R.C. (2003). Patterning the sea urchin embryo: gene regulatory networks, signaling pathways, and cellular interactions. *Current Topics in Developmental Biology* 53, 159–98.

Angerer, L.M., Oleksyn, D.W., Levine, A.M., Li, X., Klein, W.H., and Angerer, R.C. (2001). Sea urchin goosecoid function links fate specification along the animal-vegetal and oral-aboral embryonic axes. *Development* 128, 4393–404.

Berill, N.J. (1995). *The Origin of Vertebrates*. Oxford: Clarendon Press.

Bourlat, S.J., Nielsen, C., Lockyer, A.E., Littlewood, D.T., and Telford, M.J., (2003). *Xenoturbella* is a deuterostome that eats molluscs. *Nature* 424, 925–8.

Bradshaw, H.D. Jr., Otto, K.G., Frewen, B.E., McKay, J.K., and Schemske, D.W. (1998). Quantitative trait loci affecting differences in floral morphology between two species of monkeyflower (*Mimulus*). *Genetics* 149, 367–382.

Bradshaw, H.D. and Schemske, D.W. (2003). Allele substitution at a flower colour locus produces a pollinator shift in monkeyflowers. *Nature* 426, 176–8.

Budd, G.E. and Jensen, S. (2000). A critical reappraisal of the fossil record of the bilaterian phyla. *Biol Rev Camb Philos Soc* 75, 253–95.

Byrne, M. (2006). Life history diversity and the Asterinidae. *Integrative and Comparative Biology* 46, 243–54.

Byrne, M., Hart, M.W., Cerra, A., and Cisternas, P. (2003). Reproduction and larval morphology of broadcasting and viviparous species in the *Cryptasterina* species complex. *Biol Bull* 205, 285–94.

Chen, C.H., Cretekos, C.J., Rasweiler 4th, J.J., and Behringer, R.R. (2005). *Hoxd13* expression in the developing limbs of the short-tailed fruit bat, *Carollia perspicillata*. *Evolution & Development* 7, 130–41.

Cohn, M.J. and Tickle, C. (1999). Developmental basis of limblessness and axial patterning in snakes. *Nature* 399, 474–9.

Davidson, E.H., Peterson, K.J., and Cameron, R.A. (1995). Origin of bilaterian body plans: evolution of developmental regulatory mechanisms. *Science* 270, 1319–25.

De Beer, G.R. (1951). *Embryos and Ancestors*. Oxford: Clarendon Press.

Donoghue, P.C., Kouchinsky, A., Waloszek, D., Bengtson, S., Dong, X.P., Val'kov, A.K., Cunningham, J.A., and Repetski, J.E. (2006). Fossilized embryos are widespread but the record is temporally and taxonomically biased. *Evol Dev* 8, 232–8.

Duboc, V., Röttinger, E., Besnardeau, L., and Lepage, T. (2004). Nodal and BMP2/4 signaling organizes the oral-aboral axis of the sea urchin embryo. *Developmental Cell* 6, 397–410.

Duboc, V., Röttinger, E., Lapraz, F., Besnardeau, L., and Lepage, T. (2005). Left-right asymmetry in the sea urchin embryo is regulated by nodal signaling on the right side. *Developmental Cell* 9, 147–58.

Dunn, E.F., Moy, V.N., Angerer, L.M., Angerer, R.C., Morris, R.L., and Peterson, K.J. (2007). Molecular paleoecology: using gene regulatory analysis to address the origins of complex life cycles in the late Precambrian. *Evol Dev* 9, 10–24.

Ferkowicz, M.J. and Raff, R.A. (2001). Wnt gene expression in sea urchin development: heterochronies associated with the evolution of developmental mode. *Evolution & Development* 3, 24–33.

Flowers, V.L., Courteau, G.R., Poustka, A.J., Weng, W., and Venuti, J.M., (2004). Nodal/activin signaling establishes oral-aboral polarity in the early sea urchin embryo. *Developmental Dynamics* 231, 727–40.

Garstang, W. (1922). The theory of recapitulation: a critical restatement of the biogenetic law. *Journal of the Linnean Society, Zoology* 35, 81–101.

Gilbert, S.F., Loredo, G.A., Brukman, A., and Burke, A.C. (2001). Morphogenesis of the turtle shell: the development of a novel structure in tetrapod evolution. *Evolution & Development* 3, 47–58.

Gould, S.J. (1977). *Ontogeny and Phylogeny*. Cambridge, MA: Harvard University Press.

Gregory, W.K. (1951). *Evolution Emerging*. New York: Macmillan, 2 vols.

Hallgrimsson, B. and Hall, B.K. (eds.) (2005). *Variation*. Amsterdam: Elsevier.

Harada, Y., Shoguchi, E., Taguchi, S., Okai, N., Humphries, T., Tagawa, K., and Satoh, N. (2002). Conserved expression pattern of BMP-2/4 in hemichordate acorn worm and echinoderm sea cucumber embryos. *Zoological Science* 19, 1113–21.

Hart, M.W., Byrne, M., and Smith, M.J. (1997). Molecular phylogenetic analysis of life-history evolution in asterinid starfish. *Evolution* 51, 1846–59.

Heintz, A. and Stormer, L. (1937). *Relationships of the Animal Kingdom*. Oslo: Palentological Museum (wall display diagram).

Henry, J.J. and Raff, R.A. (1990). The dorsoventral axis is specified prior to first cleavage in the direct developing sea urchin *Heliocidaris erythrogramma*. *Development* 110, 875–84.
Hinman, V.F., Nguyen, A.T., Cameron, R.A., and Davidson, E.H. (2003). Developmental gene regulatory network architecture across 500 million years of echinoderm evolution. *Proceedings of the National Academy of Sciences USA*. 100, 13356–61.
Israelsson, O. and Budd, G.E. (2005). Eggs and embryos in *Xenoturbella* (phylum uncertain) are not ingested prey. *Development Genes and Evolution*. 215, 358–63.
Jågersten, G. (1972). *Evolution of the Metazoan Life Cycle*. London: Academic Press.
Jeffery, C.H. and Emlet, R.B. (2003). Macroevolutionary consequences of developmental mode in temnopleurid echinoids from the Tertiary of southern Australia. *Evolution* 57, 1031–48.
Jeffery, C.H., Emlet, R.B., and Littlewood, D.T. (2003). Phylogeny and evolution of developmental mode in temnopleurid echinoids. *Molecular Phylogenetics and Evolution* 28, 99–118.
Jenner, R.A. (2000). Evolution of animal body plans: the role of metazoan phylogeny at the interface between pattern and process. *Evolution & Development* 2, 208–21.
Kauffman, J.S. and Raff, R.A. (2003). Patterning mechanisms in the evolution of derived developmental life histories: the role of Wnt signaling in axis formation of the direct-developing sea urchin *Heliocidaris erythrogramma*. *Development Genes and Evolution* 213, 612–24.
Kirschner, M. and Gerhart, J. (1998). Evolvability. *Proc Natl Acad Sci USA* 95, 8420–27.
Kissinger, J.C. and Raff, R.A. (1998). Evolutionary changes in sites and timing of expression of actin genes in embryos of the direct- and indirect-developing sea urchins *Heliocidaris erythrogramma* and *H. tuberculata*. *Development Genes and Evolution* 208, 82–93.
Leroi, A.M. (2000). The scale independence of evolution. *Evolution & Development* 2, 67–77.
Lillie, F.R. (1898). Adaptation in cleavage. *Biological Lectures Delivered at the Marine Biological Laboratory of Wood's Hole*. Boston, MA: Ginn & Co., pp. 43–56.
Love, A.C., Andrews, M.E., and Raff, R.A. (2007). Gene expression patterns in a novel animal appendage: the sea urchin pluteus arm. *Evolution & Development* 9, 51–68.
Love, A.C. and Raff, R.A. (2003). Knowing your ancestors: themes in the history of evo-devo. *Evolution & Development* 5, 327–30.
(2006). Larval ectoderm, organizational homology, and the origins of evolutionary novelty. *J Exp Zoolog B Mol Dev Evol* 306, 18–34.
Lowe, C.J. and Wray, G.A. (1997). Radical alterations in the roles of homeobox genes during echinoderm evolution. *Nature* 389, 718–21.

McEdward, L. and Janies, D. (1997). Relationships among development, ecology, and morphology in the evolution of echinoderm larvae and life cycles. *Biological Journal of the Linnean Society* 60, 381–400.

Minsuk, S. and Raff, R.A. (2005). Co-option of an oral-aboral patterning mechanism to control left-right differentiation: the direct-developing sea urchin *Heliocidaris erythrogramma* is sinistralized, not ventralized, by $NiCl_2$. *Evolution & Development* 7, 289–300.

Müller, F. (1869). *Facts and Arguments for Darwin.* Translated from German by W.S. Dallas. London: John Murray.

Nakajima, Y., Humphreys, T., Kaneko, H., and Tagawa, K. (2004). Development and neural organization of the tornaria larva of the Hawaiian hemichordate, *Ptychodera flava. Zoological Science* 21, 69–78.

Needham, J. (1993). On the dissociability of the fundamental processes in ontogenesis. *Biol Rev Camb Philos Soc* 8, 180–223.

Nielsen, C. (1995). *Animal Evolution. Interrelationships of the Living Phyla.* Oxford University Press.

Nielsen, C. and Nørrevang, A. (1985). The trochaea theory: an example of life cycle phylogeny. In S. Conway-Morris, J.D. George, R. Gibson, and H.M. Platt (eds.), *The Origins and Relationships of Lower Invertebrates.* Oxford: Clarendon Press, pp. 297–309.

Nielsen, G., Wilson, K.A., Raff, E.C., and Raff, R.A. (2000). Novel gene expression patterns in hybrid embryos between species with different modes of development. *Evolution & Development* 2, 133–44.

Nützel, A., Lehnert, O., and Fryda, J. (2006). Origin of planktotrophy – evidence from early molluscs. *Evolution & Development* 8, 325–30.

Orr, H.A. (2005). The genetic theory of adaptation: a brief history. *Nature Reviews Genetics* 6, 119–27.

Peichel, C.L. (2005). Fishing for the secrets of vertebrate evolution in threespine sticklebacks. *Developmental Dynamics* 234, 815–23.

Peterson, K.J. (2005). Macroevolutionary interplay between planktonic larvae and benthic predators. *Geology* 33, 929–32.

Peterson, K.J., Cameron, R.A., and Davidson, E.H. (1997). Set-aside cells in maximal indirect development: evolutionary and developmental significance. *Bioessays* 19, 623–31.

Peterson, K.J., Summons, R.A., and Donoghue, P.C.J. (2007). Molecular palaeobiology. *Palaeontology* 50, 775–809.

Piatigorski, J. (2003). Crystallin genes: specialization by changes in gene regulation may precede gene duplication. *Journal of Structural and Functional Genomics* 3, 131–7.

Primus, A. (2005). Regional specification in the early embryo of the brittle star *Ophiopholis aculeata. Developmental Biology* 283, 294–309.

Raff, E.C., Popodi, E.M., Kauffman, J.S., Sly, B.J., Turner, F.R., Morris, V.B., and Raff, R.A. (2003). Regulatory punctuated equilibrium and convergence in the evolution of developmental pathways in direct-developing sea urchins. *Evolution & Development* 5, 478–93.

Raff, E.C., Popodi, E.M., Sly, B.J., Turner, F.R., Villinski, J.T., and Raff, R.A. (1999). A novel ontogenetic pathway in hybrid embryos between species with different modes of development. *Development* 126, 1937–45.

Raff, E.C., Villinski, J.T., Turner, F.R., Donoghue, P.C., and Raff, R.A. (2006). Experimental taphonomy shows the feasibility of fossil embryos. *Proceedings of the National Academy of Sciences USA* 103, 5846–51.

Raff, R.A. (1996). *The Shape of Life: Genes, Development and the Evolution of Animal Form*. University of Chicago Press.

Raff, R.A. and Love, A.C. (2004). Kowalevsky, comparative evolutionary embryology, and the intellectual lineage of evo devo. *J Exp Zoolog B Mol Dev Evol* 15, 302, 19–34.

Reisz, R.R., Scott, D., Sues, H.D., Evans, D.C., and Raath, M.A. (2005). Embryos of an early Jurassic prosauropod dinosaur and their evolutionary significance. *Science* 309, 761–4.

Richardson, M.K. and Keuck, G. (2002). Haeckel's ABC of evolution and development. *Biological Reviews* 77(4), 495–528.

Rouse, G.W. (2000). The epitome of hand waving? Larval feeding and hypotheses of metazoan phylogeny. *Evolution & Development* 2, 222–33.

Sander, K. and Schmidt-Ott, U. (2004). Evo-devo aspects of classical and molecular data in a historical perspective. *J Exp Zoolog B Mol Dev Evol* 302, 69–91.

Shapiro, M.D., Marks, M.E., Peichel, C.L., Blackman, B.K., Nereng, K.S., Jonsson, B., Schluter, D., and Kingsley, D.M. (2004). Genetic and developmental basis of evolutionary pelvic reduction in threespine sticklebacks. *Nature* 428, 717–23.

Shubin, N.H., Draeschler, E.B., and Jenkins, F.A. Jr. (2006). The pectoral fin of *Tiktaalik roseae* and the origin of the tetrapod limb. *Nature* 440, 747–49.

Slack, F. and Ruvkun, G. (1997). Temporal pattern formation by heterochronic genes. *Annual Review of Genetics* 31, 611–34.

Sly, B.J., Snoke, M.S., and Raff, R.A. (2003). Who came first – larvae or adults? Origins of bilaterian metazoan larvae. *International Journal of Developmental Biology* 47, 623–632.

Smith, A.B. (1997). Echinoderm larvae and phylogeny. *Annual Reviews of Ecology and Systematics* 28, 219–41.

Smith, M.S., Turner, F.R., and Raff, R.A. (2008). Nodal expression and heterochrony in the evolution of dorsal–ventral and left–right axes formation in the direct-developing sea urchin *Heliocidaris erythrogramma*. *J Exp Zoolog B Mol Dev Evol* 13 Aug (epub ahead of print).

Valentine, J.W. and Collins, A.G. (2000). The significance of moulting in ecdysozoan evolution. *Evolution & Development* 2, 152–6.

Villinski, J.T., Villinski, J.C., and Raff, R.A. (2002). Convergence in maternal provisioning strategy during developmental evolution of sea urchins. *Evolution* 56, 1764–75.

Wagner, G.P. (1996). Homologues, natural kinds and the evolution of modularity. *Am Zool* 36, 36–43.

Walossek, D. (1993). The upper Cambrian *Rehbachiella kinnekullensis* Müller, 1983, and the phylogeny of Branchiopoda and Crustacea. *Fossils and Strata* 32, 1–202, 54 text figures, 34 plates.

Whiteley, H.R., McCarthy, B.J., and Whiteley, A.H. (1970). Conservation of base sequences in RNA for early development of echinoderms. *Developmental Biology* 21, 216–42.

Wilson, K., Andrews, M.A., and Raff, R.A. (2005a). Dissociation of expression patterns of homeodomain transcription factors in the evolution of developmental mode in the sea urchins *Heliocidaris tuberculata* and *H. erythrogramma. Evolution & Development* 7, 401–15

Wilson, K., Andrews, M.A., Turner, F.R., and Raff, R.A. (2005b). Major regulatory factors in the evolution of development: the roles of goosecoid and Msx in the evolution of the direct-developing sea urchin *Heliocidaris erythrogramma. Evolution & Development* 7, 416–28.

Wolpert, L. (1999). From egg to adult to larva. *Evolution & Development* 1, 3–4.

Wray, G.A. (1995). Evolution of larvae and developmental modes. In L. McEdward (ed.), *Ecology of Marine Invertebrate Larvae.* Boca Raton, FL: CRC Press, pp. 413–47.

(1996). Parallel evolution of non-feeding larvae in echinoids. *Systematic Biology* 45, 308–22.

Wray, G.A. and Raff, R.A. (1989). Evolutionary modification of cell lineage in the direct-developing sea urchin *Heliocidaris erythrogramma. Developmental Biology* 132, 458–70.

(1990). Novel origins of lineage founder cells in the direct-developing sea urchin *Heliocidaris erythrogramma. Developmental Biology* 141, 41–54.

Xiao, S., Zhang, Y., and Knoll, A.H. (1998). Three-dimensional preservation of algae and animal embryos in a Neoproterozoic phosphate. *Nature* 391, 553–8.

Zigler, K.S. Raff, E.C., Popodi, E., Raff, R.A., and Lessios, H.E. (2003). Adaptive evolution of bindin in the genus *Heliocidaris* is correlated with the shift to direct development. *Evolution* 57, 2293–302.

5

A focus on both form and function in examining selection versus constraint

PAUL M. BRAKEFIELD

One of the challenges for a modern integrative biology is to more fully understand why assemblages of related species occupy morphospace in the way they do. An integration of analyses of both form and function is required to achieve this goal. The burgeoning understanding of development in emerging model species is providing evolutionary biologists with the potential to explore contributions to the evolution of patterns in morphospace by the processes involved in making variation in forms as well as those resulting from their performance in natural environments. Such an approach is illustrated here by drawing on work on butterfly eyespot patterns and scaling relationships. Relevant studies are being performed in some other groups of animals (see Brakefield 2006), and progress is also being made in plants (see Langlade *et al.* 2005; Niklas, ch. 3 in this volume).

Patterns in morphospace have always fascinated biologists, especially when they are known to reflect adaptive radiation among new ecological niches. In such examples, while there can be no doubt that natural selection plays a major role in shaping the evolution of morphological diversity and disparity, it is by no means clear how much the processes involved in generating variation in the phenotype to be screened by the natural selection also contribute. Thus, the extent to which evolution by natural selection of adaptations to local environments is compromised or biased by the genetical and developmental origins of phenotypic variation remains an open issue. For example, genetic channelling

I would like to thank the organizers of the symposium for their invitation. David Jablonsky made extremely useful comments on an early draft, especially in relation to the work on snail shells. It would not have been possible to write this chapter without the diversity of contributions made by members of the *Bicyclus* laboratory in Leiden.

arising from genetic correlations and pleiotropy among traits could result in the clustering of species along so-called axes of least resistance (e.g. Blows and Hoffmann 2005; Schluter 1996; Via and Lande 1985). Similarly, a bias in the production of particular phenotypes emerging from properties of the development system may also result in such clustering (e.g. Arthur 2001; Maynard Smith *et al.* 1985; Wagner and Altenberg 1996). Although such notions about genetic channelling or developmental bias or drive are not new, their contribution, alongside natural selection, to patterns of evolution in the fossil record or among extant taxa is unclear (Brakefield 2006).

Here, I will illustrate through our research on butterfly wings how the emergence of a broad Evo Devo or integrative evolutionary biology is beginning to provide the impulse to attempt more experimental analyses of intrinsic versus extrinsic processes in evolution (Allen *et al.* 2008; Brakefield 2006; Brakefield and Roskam 2006). We seek to examine both the intrinsic roles of genetic channelling and developmental bias, and the extrinsic processes of natural selection in shaping patterns of morphological diversity among related taxa. This approach is based on artificial selection experiments, using an emerging model organism that can be rapidly mass-reared in the laboratory, to test ideas about how patterns of morphological evolution could be influenced by genetic correlations and developmental bias. Ultimately, such analyses of the potential involvement of intrinsic factors can be combined with different ways of examining how natural selection influences the phenotypic variances for the same traits. The latter may involve experiments with the same model organism and comparative analyses of morphological patterns of diversity and disparity in the whole lineage.

Evolvability is a more encompassing term than either genetic channelling or developmental bias (e.g. Dawkins 1989; Kauffmann 1985; Kirschner and Gerhart 1998; Wagner 2005; Wagner and Altenberg 1996). A straightforward definition of this term is the capacity to adapt to changing conditions; it can be applied at different levels – for example, to a specific trait or complex form, a population, or a lineage. Both the processes involved in generating form, as well as the ways in which different traits contribute to function, are components of evolvability. Ideas about how the evolvability of sets of interacting traits – whether involving genetic channelling, developmental bias or other phenomena – can change through time are all keys to accounting for patterns of occupancy of morphospace. All these types of concepts are interrelated, and more experimental data from new systems that can be analysed at different levels of biological organization are required to move ahead in our overall understanding of the contributions of both intrinsic and extrinsic processes to patterns of morphological variation.

CASE STUDIES ON SNAIL SHAPE AND SIZE

A series of studies that is especially illustrative of these types of issues was initiated by the ecological geneticist Arthur Cain working in Oxford and Liverpool, and the palaeontologist David Raup in Chicago. Cain is perhaps best remembered for his work on the ecological genetics of shell colour and banding polymorphism in *Cepaea nemoralis* and *C. hortensis* (Cain and Sheppard 1954). However, he was also fascinated by the diversity and disparity in the shape and size of the shells within terrestrial faunas of gastropod molluscs (Cain 1977, 1981). He made plots of variation among species in the spire index (essentially shell height times breadth). He then attempted to relate the patterns in morphological diversity in form to different ecological lifestyles or, in general terms, to function. Thus, some morphotypes of the snails lived primarily beneath stones, while others were found mainly on tree surfaces.

Although Raup is perhaps best known for his work on major episodes of biological extinction of marine families, he also performed a highly influential analysis of how species of gastropod snails fill plots of potential morphological space for shell form, as described by three parameters of growth (Raup 1966, 1967; see also fig. 5.1). Such maps of species occurrence in morphospace (the 'snail-cube' in Raup's case) generally emphasize how large parts of the potential space are not filled. This may suggest hypotheses about the involvement of natural selection in shaping broad features of the observed patterns in morphological diversity; again, particular morphotypes are likely to be best suited for particular lifestyles. In the case of the snail-cube, explanations also involved regions with no occupancy due to mechanical reasons (for overviews, see McGhee 1999, 2007). However, such descriptions of diversity in form in themselves usually provide little information relevant to exploring whether the processes that generate variation in the morphologies are also involved in their evolution (but see Niklas, ch. 3 in this volume).

Since these early works, snails have continued to be a target for describing patterns of diversity and disparity in form within both terrestrial and marine faunas (e.g. Cowie 1995; Gittenberger 2006). More recent work has confirmed that even within species, phenotypic variation

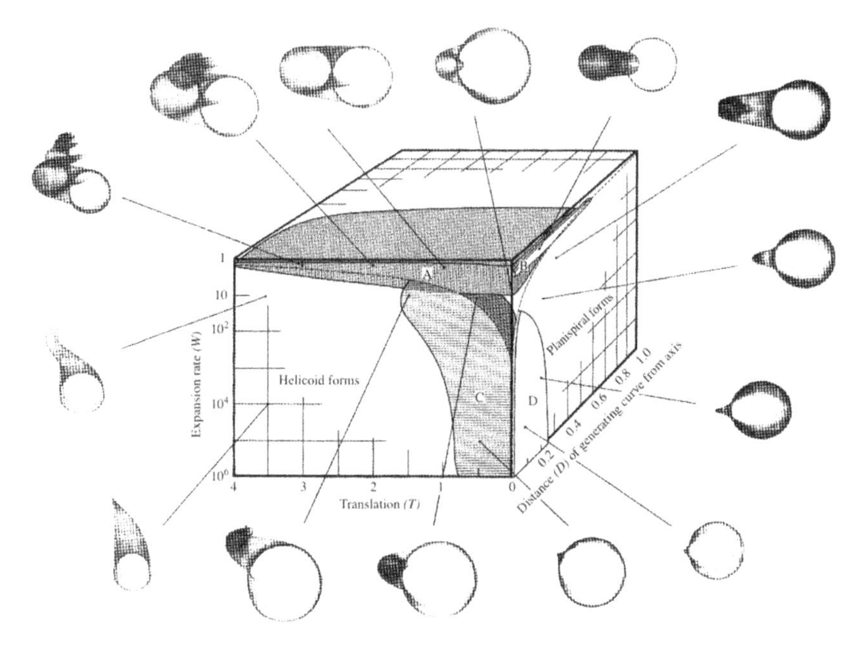

Figure 5.1 Morphospace illustrated by the 'snail cube' of David Raup. The three-dimensional cube describes a set of possible shell shapes, whereas only the four regions marked A, B, C, and D are occupied by species of snail. All other regions represent theoretically possible shell shapes but ones which are unrealized by snails in nature. Fourteen possible shell shapes, as drawn by computer, are arranged around the periphery. From Raup 1966 and Ridley 2003.

in the shape and size of snail shells is influenced by natural selection (e.g. Boulding and Hay 1993; Cook 1990; Goodfriend 1986). Other researchers have explored ideas about constructional, geometrical, or architectural constraints on shell morphology (e.g. Stone 1996), and performed some clever tests of certain modes of constraints – for example, by analysing patterns of transition in the fossil record (Wagner 1995). Moreover, this group of organisms remains one of the most exciting for establishing patterns of diversity and disparity through geological time (e.g. Ciampaglio *et al.* 2001; McClain 2005; McClain *et al.* 2004; Wagner and Erwin 2006). Ecology, genetics, development, and macroevolution in the fossil record can truly be thought of as intersecting at this level (Jablonski 2005). However, to be successful in linking the contributions of intrinsic and extrinsic processes, the parameters of growth identified theoretically by Raup will need to be understood in terms of genetic variation and specific developmental mechanisms. The studies required to achieve this will probably remain a major challenge in gastropods for the foreseeable future, but some other systems are starting to provide the ability to work experimentally at the genetic and developmental levels, as well as in terms of natural selection and adaptive evolution.

CASE STUDIES ON BUTTERFLY WINGS

Evo Devo is extending the work on the few model organisms of developmental biology – worms, flies, frogs, and mice – to examine more subtle variation in morphology characteristic of differences among closely related species. This will reveal the extent to which genes with crucial functions in development (including in embryos) harbour segregating variation within natural populations that contributes to the developmental basis of morphological variation. The spectacular diversity of butterfly wing patterns is yielding a series of such studies (Beldade and Brakefield 2002; Joron *et al.* 2006). Research on *Drosophila* flies has provided avenues into understanding the developmental genetics of butterfly wings. Thus, developmental biologists working with flies and using stocks from mutagenic screens have revealed the 'toolkit' pathways of developmental genes which control the development of wings and of their constituent structures, including trachea (veins) and bristles (see Carroll *et al.* 2005). The same pathways are responsible for the development of homologous structures in butterflies. Work has begun to show how components of these pathways have been co-opted, and often elaborated upon, to yield key, novel structures in butterfly wings, including the pigmented scale cells (from bristles and the *Achete-scute* pathway; Gallant *et al.* 1998) and eyespot pattern elements (co-option of genes such as *Distal-less, engrailed*, and *Spalt*; Brunetti *et al.* 2001; Carroll *et al.* 1994; Keys *et al.* 1999; Saenko *et al.* 2008). In contrast to many species of *Drosophila* flies, butterflies are often more amenable in the field, where they may have a fascinating biology, such as seasonal forms in *Bicyclus* or mimicry in *Heliconius*. Combining insights from Evo Devo studies about how form is generated with this ability to open up the functional significance of the same forms will become increasingly important.

The colour patterns on the scale-covered wings of butterflies in the family Nymphalidae are made up of combinations of different pattern elements, including colour bands, stripes, and marginal ocelli or eyespots (Nijhout 1991). A reconstruction of a nymphalid 'ground plan' shows repeated series of the different types of element arranged in anterior-posterior columns on each surface of the fore- and hindwings. The repeated series of a particular element can be considered as a module. Each surface of a wing is subdivided by wing veins into wing cells, each of which has its own combination of single repeats of the different pattern elements (see fig. 5.2). The development of one set of such elements, the marginal eyespots, is becoming understood both in terms of cell–cell signalling mechanisms and candidate genetic pathways (Beldade and Brakefield 2002). Essentially, understanding development involves discovering how different populations of epithelial cells in the wings – the scale cells to be – gain information during wing growth in the late larval and early pupal stages, and thus become fated to lay down different colour pigments just before adult eclosion. This understanding of morphogenesis and pattern formation can now be used alongside artificial selection experiments targeted on the pattern of eyespots to explore whether evolvability and the mechanisms involved in generating phenotypic variation influence evolutionary trajectories.

SERIALLY REPEATED ELEMENTS AND ARTIFICIAL SELECTION EXPERIMENTS ON EYESPOTS

Eyespots in *Bicyclus* as well as in other butterflies and moths (together, the Lepidoptera) are known to function both in interactions with predators and during mate choice (Brakefield and Frankino 2006; Breuker and Brakefield 2002; Robertson and Monteiro 2005; Stevens 2005). Thus, both natural selection and sexual selection are relevant to understanding functional differences in eyespot patterns among species. The eyespots of *B. anynana* are all formed in the late larval stage and the early pupa by the same developmental pathway (Beldade and Brakefield 2002; Brunetti *et al.* 2001; Reed and Serfas 2004). Transplantation experiments performed in early pupae show they are formed around groups of organizing cells called foci; transplanting an eyespot focus to a novel site in the pupal wing yields an ectopic eyespot around the grafted tissue in the adult wing (French and Brakefield 1995). Establishment of the foci occurs in late larvae, and then in early pupae each focus sets up a gradient in the surrounding epithelial cells, presumably via one or more diffusible morphogens. These cells then respond to the gradient of the signal and, depending in some way on concentration, become fated to synthesize a particular colour pigment just before emergence of the adult. We also know that a large eyespot is produced by a stronger signalling focus in the early pupa than occurs in a small eyespot (Monteiro *et al.* 1994).

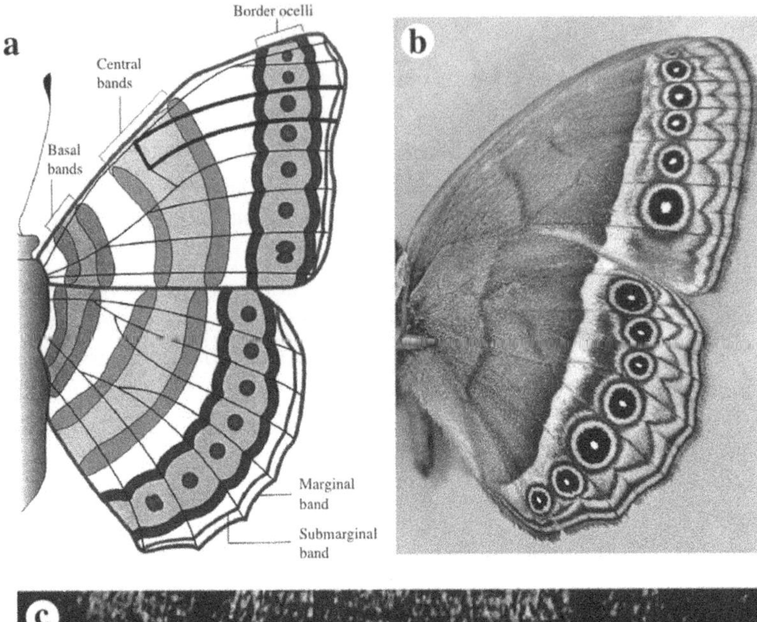

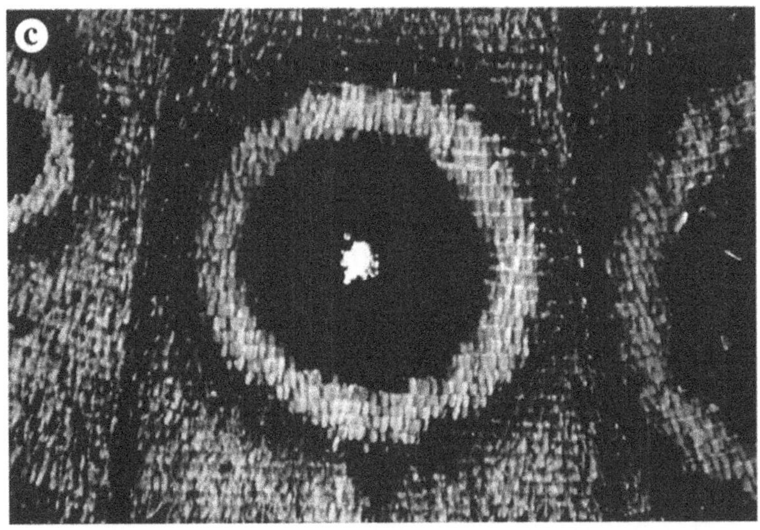

Figure 5.2 Butterfly wing patterns and eyespots. (a) The 'nymphalid ground plan' shown on a forewing and hindwing. According to this idealized ground plan, the types of pattern elements such as the border ocelli or eyespots are organized in parallel series. Homologous pattern elements are serially repeated in individual compartments bordered by veins (the boundaries of one such compartment, known as a wing cell, are illustrated by thickened lines). (b) Ventral wing surfaces of *Mycalesis horsfeldi*, showing a pattern closely similar to the ground plan. *Mycalesis* is a speciose genus closely related to *Bicyclus*, but distributed throughout Asia. (c) Detail of an eyespot with concentric rings of differing colour on the ventral hindwing of *B. anynana*. Note that the individual epithelial scale cells are visible as rows of small rectangles, especially in the gold ring. (a) redrawn from Nijhout 1991; (b) photo by J.C. Roskam; (c) from Brakefield *et al.* 1996.

The eyespots on both the dorsal and ventral wing surfaces of *B. anynana* express the same developmental genes, and at comparable stages in eyespot formation (Beldade *et al.* 2002, 2005; Brunetti *et al.* 2001; Reed and Serfas 2004). These genes include *Distal-less*, *hedgehog*, *engrailed*, *Spalt*, and *Notch*. Typically, mutant alleles established in laboratory stocks also affect all eyespots. Moreover, artificial selection experiments targeted on the size or colour of a single eyespot yield highly correlated responses for the target trait in other eyespots, especially in those on the same wing surface. The shared morphogenesis, both in terms of genetic variation and developmental mechanisms, led us to design experiments with *B. anynana* to examine the potential developmental flexibility of the repeated eyespot elements to evolve in different directions in trait space or developmental morphospace (Brakefield 1998).

The experiments explored whether a form with a pair of eyespots in which one was smaller and the other larger could be produced as readily by artificial selection as one in which both eyespots were either larger or smaller than in the wild type. Predictions based on the strong correlated responses observed when selection was targeted on a single eyespot (Monteiro *et al.* 1994) were that a change in morphology would be more limited in rate or extent for the pattern in which eyespots were selected in opposite directions (Brakefield 1998). We targeted the wild-type pattern for the forewing of the butterfly that shows two eyespots: a small anterior one and a larger posterior one (Beldade *et al.* 2002). Replicated lines were established from the same founder population and selected towards each of the four corners of trait space – that is, along the 'coupled' axis towards either two small or two large eyespots, as is consistent with the shared genetics and development; and along the 'uncoupled' axis where the two eyespots are selected in opposite directions. This latter axis, therefore, represents one orthogonal to the proposed genetic line of least resistance and the plane of developmental bias (see fig. 5.3).

Artificial selection occurred over twenty-five generations (Beldade *et al.* 2002). As expected, selection either 'up' or 'down' the 'coupled' axis of shared development produced rapid responses, with butterflies eventually either having no eyespots or two very large ones. These morphologies are completely different to any present in the base population and are therefore highly novel. However, populations along the alternative 'uncoupled' axis orthogonal to that reflecting shared development also responded well to selection, eventually producing phenotypes in which one eyespot was very large and the other absent or very

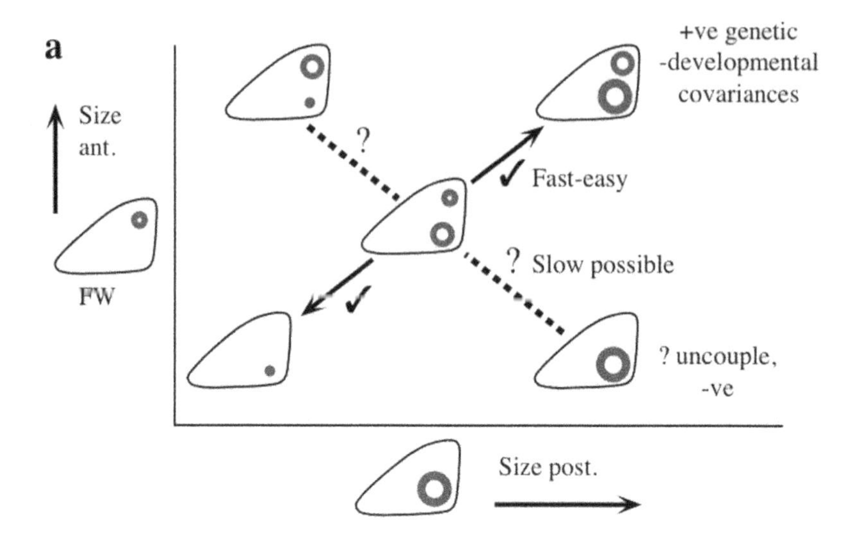

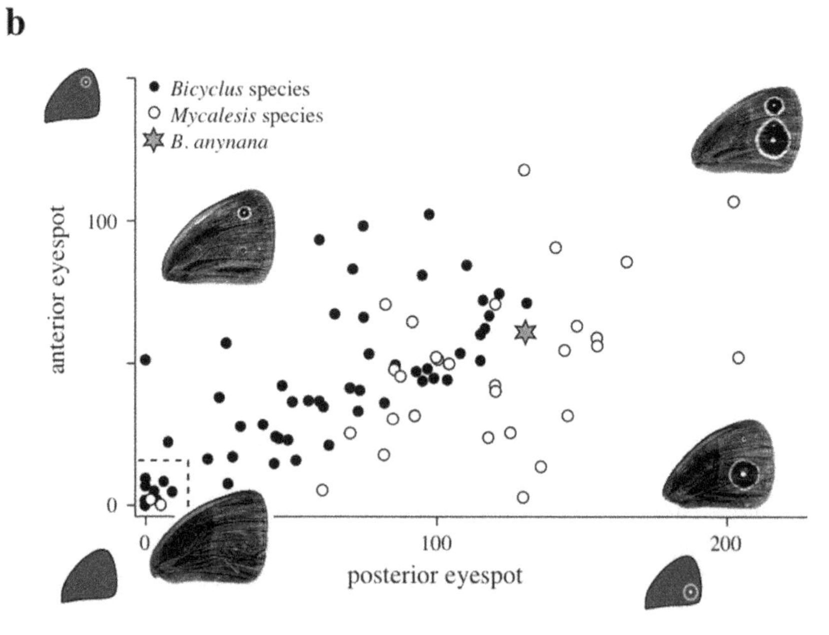

Figure 5.3 Analysis of morphospace for two eyespot traits in *Bicyclus* and *Mycalesis* butterflies. (a) Diagram illustrating the design of an artificial selection experiment on *Bicyclus anynana* to explore the rate of response to selection in different directions through morphospace for the relative size of the two dorsal forewing eyespots, starting from a base population with the wild-type pattern illustrated in the central figure. Previous work had shown that the leading diagonal represents an axis of shared genetical and developmental mechanisms

small. Again, these represent highly novel phenotypes. Thus, this pattern in the relative size of the two eyespots behaved in a developmentally flexible manner, with a high evolvability in all directions through morphospace (Beldade *et al.* 2002). There is no reason to imagine that other directions of change, such as ones in which one eyespot changes and the other does not, would be any more resistant to change if under the influence of equally intense and targeted selection. We suggested that this capacity for independent evolution was the product of a long legacy of natural selection and evolutionary tinkering, leading to morphological diversity among species and to corresponding evolvabilities for eyespot size at different sites in the wings (Beldade *et al.* 2003).

Current work is focused on a different eyespot trait, namely colour composition. This work is also based on quantitative variation in an outcrossed stock population of *B. anynana*. Thus, there is some variability with respect to whether particular eyespots have a comparatively narrow outer gold ring or a broader one, relative to the size of the inner black disc which surrounds the white, central pupil of the eyespot. Work on a single eyespot had already demonstrated that this trait has a similarly high heritability as eyespot size, and that positive genetic correlations occur among eyespots (Monteiro *et al.* 1997). There is, however, only a very low genetic correlation between eyespot colour and eyespot size, indicating that different sets of alleles modulate these traits. Again, artificial selection along the 'coupled' axis for a specific pair of eyespots

Caption for Figure 5.3 *(continued)*

for the two eyespots, along which the pattern could be readily changed. The prediction was that morphological change would be more limited along the other diagonal, where selection was targeted to uncouple the size of the two eyespots. (b) Responses in morphospace for the two dorsal forewing eyespots in the selection experiments in *B. anynana* compared to the variation among species of this African genus and of the closely related Asian genus *Mycalesis*. The four images of the forewing to each corner of the morphospace are representative examples of the wing pattern after twenty-five generations of artificial selection in *B. anynana* (Beldade *et al.* 2002). The wild-type pattern for this species is depicted by the star. The four wings are placed in roughly their correct positions in the depicted two-trait space. Circles show the positions of the mean patterns of the size of the homologous eyespots for different species of *Bicyclus* and *Mycalesis* (closed and open symbols, respectively). The dotted square encloses species for which both eyespots are very small or absent, and frequently difficult to measure. (a) from Brakefield 1998; (b) from Brakefield and Roskam 2006.

rapidly yields novel morphologies in which both targeted eyespots have narrower or broader gold rings (as well as any flanking eyespots). In contrast to eyespot size, though, morphological change is much more strongly limited along the 'uncoupling' axis, in which eyespots are selected in opposite directions in morphospace (Allen *et al.* 2008).

These experiments using artificial selection on the evolution of the eyespot pattern explore the potential roles of flexibility in genetic variation and developmental mechanisms in shaping patterns of evolutionary change. However, they do not directly examine the evolvability of form in the context of the functional issues of natural selection on the performance of the forms. Although we know from other studies that eyespot size in species of *Bicyclus* can be a target for both natural and sexual selection (Brakefield and Frankino 2006; Breuker and Brakefield 2002; Robertson and Monteiro 2005), further work remains to be done with the different forms yielded by artificial selection in our experiments. We are starting to examine whether the properties of the responses to artificial selection for different eyespot traits are reflected in the patterns of morphological disparity across species in the lineage (Brakefield and Roskam 2006). It is, however, necessary to consider potential challenges in extrapolating from artificial selection experiments that involve a single outcrossed laboratory stock of one model species to morphologies found in related species.

SOME LIMITATIONS TO USING ARTIFICIAL SELECTION

Initially, we can consider how far the responses observed in artificial selection experiments can inform us about the role of development in generating phenotypic variation in morphology that can then be a target for natural selection in the ecological arena. The answer is probably not very far without any further sources of information. However, phenotypes (and genotypes) yielded by artificial selection can be used to explore the morphological changes in terms of the mechanisms of development; follow-up studies can identify how variability in development or differing 'options for change' have yielded the observed changes in form. It is also worth emphasizing that because artificial selection experiments usually target standing genetic variation derived from one or more natural populations, they can sometimes provide more relevant information about the (short-term) potential for phenotypic change in nature than descriptions of variation based on mutagenic screens.

Combining artificial selection experiments on a complex morphology with insights about the developmental and genetic options for change will inform us about the potential for responses to short-term natural selection in different directions of trait space. Furthermore, matching experimental studies using a model species in the laboratory with a more comparative descriptive approach for the whole lineage will reveal the extent to which the responses in short-term selection experiments may inform about patterns of evolutionary change over long timescales. An exciting prospect of such integrative approaches for amenable systems will be the ability to compare patterns of evolvability and of evolutionary change across different traits making up complex morphologies. This will in turn provide more general insights about how the properties of change in developmental processes contribute to observed patterns of evolution and the occupancy of morphological space.

Thus, artificial selection in certain organisms can be a powerful tool for uncovering potential sources of bias intrinsic to making the phenotype in the processes of evolutionary change. However, if artificial selection is to help inform about the adaptive evolution of complex morphologies, the sources of variation targeted in such experiments must have parallels with the ways in which phenotypic variation is generated in the wild, and thus with the responses to natural selection in different environments. Revealing the extent to which phenotypes yielded by artificial selection in a laboratory model organism resemble those of related species in terms of the underlying genetic and developmental changes is an additional exciting challenge for the future. Artificial selection can be targeted very specifically on particular traits, and is likely in many cases to be less influenced by pleiotropic effects on fitness than are responses to natural selection. The relevance to evolution in the wild is then likely to depend on the extent to which variation segregating somewhere within natural populations of a species provides the bases for the differences observed among related species.

Species frequently exhibit substantial spatial and temporal variability among different local environments (e.g. Thompson 2005), and they often have leaky genetic bounds. Perhaps while some, or even many, of the alleles that differentiate related species will prove to be 'private' or diagnostic at any one time point, the associated genes and genetic pathways, as well as the types of developmental changes, will have much more in common. In other words, while a specific allele underlying a difference in a morphological trait between two related species may often not occur in one of the species and be fixed in the other, Evo Devo studies of how the phenotype is generated in a model species will reveal many of the genes crucial to understanding the species-level differences. The privacy at an allelic level may in time also be found to apply more especially to traits connected directly to speciation and reproductive isolation, and less to those traits more specific to adaptive responses to environmental gradients. Thus, standing genetic variation within species for phenotypic variation in traits characteristic of such adaptive responses, such as eyespot pattern traits in *Bicyclus* butterflies, may be more likely to include examples of alleles that underlie differences in form among related species. It will also be fascinating to determine the complexity of differences in genetic architecture for morphological traits across related species, both in terms of the numbers of underlying alleles and the size distribution of their individual effects. The answers to such issues are likely to depend on the traits and the range of environments concerned, and, in particular, on the extent to which such traits exhibit pleiotropic interactions with other traits, including those involved in the evolution of reproductive isolation.

THE EVOLUTION OF ALLOMETRY INVOLVING BUTTERFLY WINGS

In addition to these types of issues, interpretations about how the eyespot phenotypes generated by the artificial selection experiments in *B. anynana* affect performance and fitness are indirect and flow from other observations, sometimes in other species (e.g. Stevens 2005). For example, cohort analyses performed in the field in Malawi with seasonal forms in *B. safitza* have demonstrated a strong selective disadvantage for butterflies with conspicuous eyespots in the dry season (fig. 5.4). An additional set of experiments on allometric growth of the wings of *B. anynana*, however, has begun to study both the potential of artificial selection to yield novel forms in the laboratory, and the consequences of the morphological changes produced in the same selection lines for reproductive success under conditions close to those in nature. A close examination of most groups of related species will reveal the substantial contribution made to patterns of morphological diversity by evolution on the scaling proportions of different body parts of structures. These are established through the evolution of allometry, and thus to changes in the differential rate of growth of different parts of dimensions in the form of organisms.

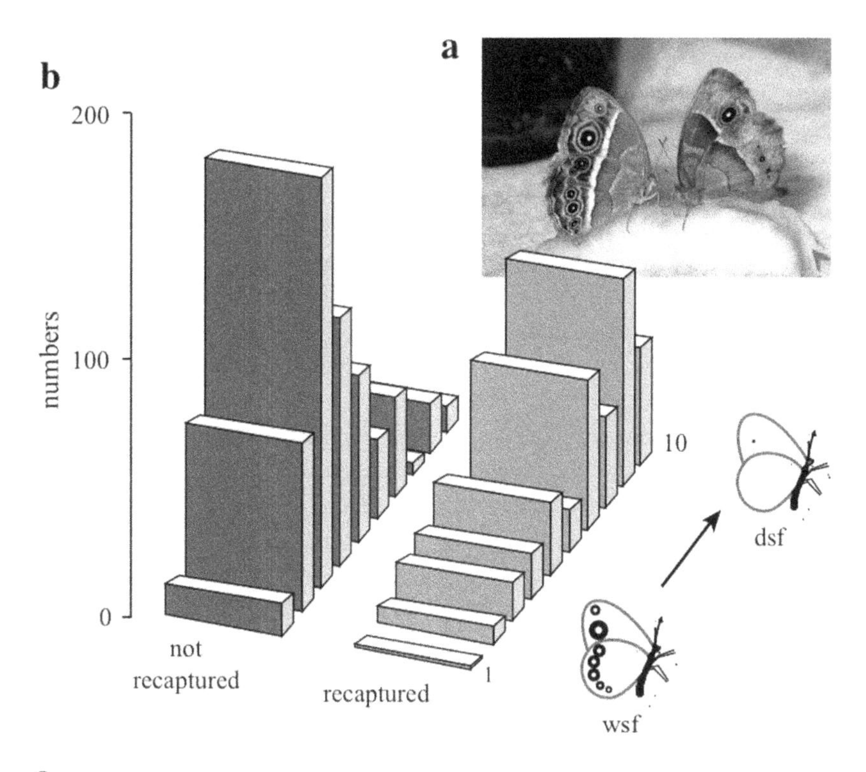

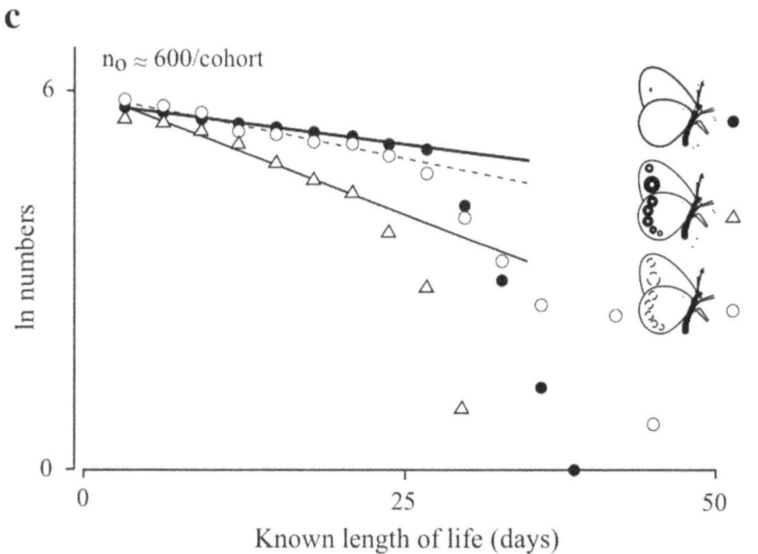

Figure 5.4 Phenotypic plasticity and natural selection in *Bicyclus safitza* butterflies. (a) Seasonal polyphenism illustrated by specimens of the wet season form (wsf) and dry season form (dsf), to the left and right, respectively.

In a parallel way to the evolution of patterns in repeated series of eyespots, diversification in the patterns of relative growth of different appendages in butterflies must involve some uncoupling of traits that originally shared all of their genetic pathways and developmental mechanisms. Both the forewings and the hindwings of butterflies express many of the same developmental genes at the same stages of development, although certain genes, including *Ultrabithorax* (*Ubx*), can provide identity to these two pairs of wings (Weatherbee *et al.* 1999). The wings also show closely congruent dynamics of growth during development. In spite of this high degree of shared development, many species of butterfly, even when rather closely related, can be highly divergent in the size of the forewings relative to the hindwings. A similar argument can be made for the size of both pairs of wings relative to the body, or 'wing-loading', which is known to have ecological consequences – for example, in powers of dispersal. Patterns of divergence among species contrast with a typically very low variation in such scaling relationships within a population of any particular species. Again, artificial selection in the

Caption for Figure 5.4 *(continued)*

The ventral wing surfaces are displayed while the butterflies are feeding on banana fruit. When resting inactively for a long time, individuals of the dsf withdraw their forewings a little between their hindwings, thus hiding the forewing posterior eyespot. These two butterflies are sisters. Although they are of similar genotype, they differ in wing pattern because whereas the wsf individual was reared at 27 °C, its dsf sister was switched to 17 °C in the late larval stage, prior to pupation. These forms occur in alternating generations in wet and dry seasons that are also warm or cool, respectively. The dorsal wing surfaces do not differ between the seasonal forms. (b) and (c) Natural selection on ventral eyespot size in *Bicyclus safitza* butterflies, as demonstrated by field experiments in the dry season in Africa, when butterflies with a highly effective camouflage and a cryptic wing pattern are expected to be more likely to survive from predators. (b) Probability of recapture over a grid of forest traps for releases of reared butterflies of ten phenotypic classes, ranging from the extreme wet season form with very large eyespots to the extreme dry season form with no eyespots. The wsf shows much higher mortality. (c) Survivorship curves for releases of about 1,800 dsf butterflies collected in another forest and divided among three treatments: unpainted controls, painted using felt-tip pens with conspicuous eyespots, and painted with inconspicuous eyespots. Butterflies with painted, conspicuous eyespots show a dramatically higher mortality, consistent with the eyespots making them easier to find by predators in the dry season. Lines show periods of age-independent survival for each cohort. From Brakefield and Frankino 2008.

model species *B. anynana* is exploring the potential flexibility in short-term responses in these allometries within this species.

The application of artificial selection in laboratory populations has resulted in divergence in the scaling relationship for wing size relative to body size, and for forewing to hindwing size (Frankino *et al.* 2005, 2007), in each case producing novel morphologies or forms relative to those in the baseline population. Following selection, the populations with divergent scaling relationships were each crossed to produce single populations with wide phenotypic variance for each scaling relationship. These latter populations were then used to compare the mating success for males showing changed allometry with males of wild-type allometry. We used competition experiments with marked males in a spacious tropical greenhouse, combined with a technique involving recapturing females after mating and tracking their male partners via the transfer of florescent dusts of different colours during copulation (cf. Joron and Brakefield 2003). For each scaling relationship, the wild-type males had substantially higher mating success than either of the divergent phenotypes (e.g. fig. 5.5). These studies provide support for the occurrence

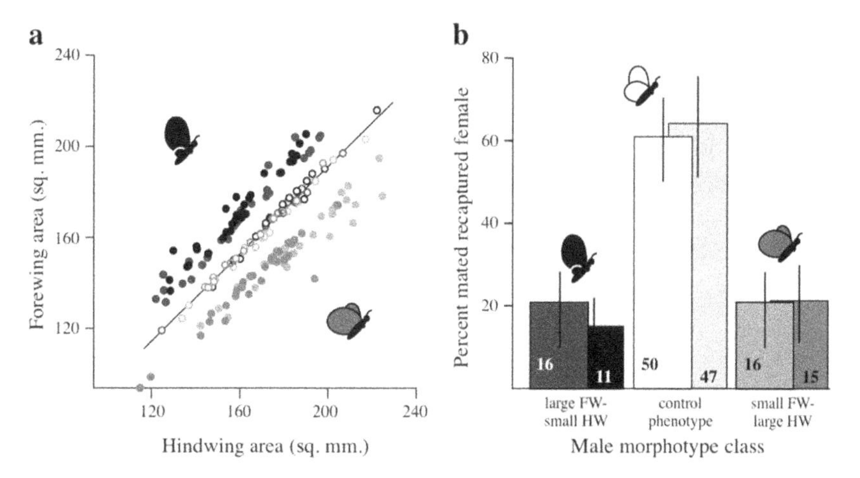

Figure 5.5 The relative mating success of male *Bicyclus anynana* with differing wing allometries. (a) Phenotypes of the groups of males used in the competition experiments, as produced by selecting individuals from a large hybrid population from a cross of two lines after artificial selection in the directions of the cartoons: males with comparatively large forewings (dark circles); males of wild-type allometry (open circles); and males with large hindwings (grey circles). (b) Measurements of the relative mating success of these males in competition for wild-type females in free flight in a spacious tropical greenhouse. From Frankino *et al.* 2007.

of strong stabilizing selection within species in specific environments, but also suggest the necessary evolvability to account for the evolution of diversity in the form of scaling relationships for species in environments with differing functional requirements and patterns of natural selection.

PROSPECTS

This analysis of allometry in *B. anynana* provides an example of the scope of an integrative evo-devo with amenable systems (see Brakefield and French 2006) to examine the extent to which the processes of generating variation in form contribute alongside the screening of functional performance of different forms by natural selection to patterns in the occupancy of morphospace. The studies on eyespot patterns in the same species can also be extended in a similar way by appropriate experiments on natural selection, using the novel forms yielded by the artificial selection in different directions of morphospace. It will then be rewarding to extend the results of these types of approaches to analysing form and function in single-model species, such as *B. anynana*, to patterns of diversity and disparity in morphology in whole lineages (see fig. 5.3b). We will then be in a better position to assess the importance of concepts such as genetic channelling and developmental bias, as well as natural selection, to shaping the evolution of form and function. In conclusion, only when data analysing function can be integrated with knowledge about the development of the relevant forms will the relative contributions of intrinsic and extrinsic processes to the evolution of occupancy of morphospace be clear.

REFERENCES

Allen, C., Beldade, P., Zwaan, B. J., and Brakefield, P.M. (2008). Differences in the selection response of serially repeated color pattern characters: standing variation, development, and evolution. *BMC Evolutionary Biology* 8, 94.

Arthur, W. (2001). Developmental drive: an important determinant of the direction of phenotypic evolution. *Evol Dev* 3, 271–8.

Beldade, P. and Brakefield, P.M. (2002). The genetics and evo-devo of butterfly wing patterns. *Nat Rev Genet* 3, 442–52.

Beldade, P., Brakefield, P.M., and Long, A.D. (2005). Generating phenotypic variation: prospects from 'evo-devo' research on *Bicyclus anynana* wing patterns. *Evol Dev* 7, 101–7.

Beldade, P., Koops, K., and Brakefield, P.M. (2002). Developmental constraints versus flexibility in morphological evolution. *Nature* 416, 844–7.
(2003). Modularity, individuality, and evo-devo in butterfly wings. *Proc Natl Acad Sci USA* 99, 14262–7.
Blows, M.W. and Hoffmann, A.A. (2005). A reassessment of genetic limits to evolutionary change. *Ecology* 86, 1371–84.
Boulding, E.G. and Hay, T.K. (1993). Quantitative genetics of shell form of an intertidal snail: constraints on short-term response to selection. *Evolution* 47, 576–92.
Brakefield, P.M. (1998). The evolution–development interface and advances with the eyespot patterns of *Bicyclus* butterflies. *Heredity* 80, 265–72.
(2006). Evo-devo and constraints on selection. *Trends Ecol Evol* 21, 362–8.
Brakefield, P.M. and Frankino, W.A. (2008). Polyphenisms in Lepidoptera: multidisciplinary approaches to studies of evolution. In D.W. Whitman and T.N. Ananthakrishnan (eds.), *Phenotypic Plasticity of Insects: Mechanisms and Consequences*. Plymouth: Science Publishers, Inc., pp.121–51.
Brakefield, P.M. and French, V. (2006). Evo-devo focus issue. *Heredity* 97, 137–8.
Brakefield, P.M. and Roskam, J.C. (2006). Exploring evolutionary constraints is a task for an integrative evolutionary biology. *Am Nat* 168, S4–S13.
Breuker, C.J. and Brakefield, P.M. (2002). Female choice depends on size but not symmetry of dorsal eyespots in the butterfly *Bicyclus anynana. Proc R Soc Lond B* 269, 1233–9.
Brunetti, C.R., Selegue, J.E., Monteiro, A., French, V., Brakefield, P.M., and Carroll, S.B. (2001). The generation and diversification of butterfly eyespot color patterns. *Curr Biol* 11, 1578–85.
Cain, A.J. (1977). Variation in the spire index of some coiled gastropod shells, and its evolutionary significance. *Philos Trans R Soc Lond B Biol Sci* 277, 333–424.
(1981). Variation in shell shape and size of helicid snails in relation to other Pulmonates in faunas of the Palaeacrtic region. *Malacologia* 21, 149–76.
Cain, A.J. and Sheppard, P.M. (1954). Natural selection in *Cepaea. Genetics* 39, 89–116.
Carroll, S.B., Gates, J., Keys, D., Paddock, S.W., Panganiban, G.F., Selegue, J., and Williams, J.A. (1994). Pattern formation and eyespot determination in butterfly wings. *Science* 265, 109–14.
Carroll, S.B., Grenier, J.K., and Weatherbee, S.D. (2005). *From DNA to Diversity: Molecular Genetics and the Evolution of Animal Design*. Oxford: Blackwell Scientific.
Ciampaglio, C.N., Kemp, M., and McShea, D.W. (2001). *Paleobiology* 27, 695–715.
Cook, L.M. (1990). Differences in shell properties between morphs of *Littoraria pallescens*. *Hydrobiologia* 193, 217–21.
Cowie, R.H. (1995). Variation in species diversity and shell shape in Hawaiian land snails: in situ speciation and ecological relationships. *Evolution* 49, 1191–202.

Dawkins, R. (1989). The evolution of evolvability. In C. Langton (ed.), *Artificial Life*. Redwood City, CA: Addison-Wesley, pp.201–20.

Frankino, W.A., Zwaan, B.J., Stern, D.L., and Brakefield, P.M. (2005). Natural selection and developmental constraints in the evolution of allometries. *Science* 307, 718–20.

(2007). Internal and external constraints in the evolution of allometries among morphological traits in a butterfly. *Evolution* 61(12), 2958–7.

French, V. and Brakefield, P.M. (1995). Eyespot development on butterfly wings: the focal signal. *Dev Biol* 168, 112–23.

Galant, R., Skeath, J.B., Paddock, S., Lewis, D.L., and Carroll, S.B. (1998) Expression of an *achaete-scute* homolog during butterfly scale development reveals the homology of insect scales and sensory bristles. *Curr Biol* 8, 807–13.

Gittenberger, A. (2006). The evolutionary history of parasitic gastropods and their coral hosts in the Indo-Pacific. Ph.D. thesis, Leiden University, The Netherlands.

Goodfriend, G.A. (1986). Variation in land snail shell form and size and its causes: a review. *Syst Zool* 35, 204–23.

Jablonski, D. (2005). Evolutionary innovations in the fossil record: the intersection of ecology, development and macroevolution. *J Exp Zoolog B Mol Dev Evol* 304B, 504–19.

Joron, M. and Brakefield, P.M. (2003). Captivity masks inbreeding effects on male mating success in butterflies. *Nature* 424, 191–4.

Joron, M., Jiggins, C.D., Papanicolaou, A., and McMillan, W.O. (2006). *Heliconius* wing patterns: an evo-devo model for understanding phenotypic diversity. *Heredity* 97, 157–67.

Kauffmann, S.A. (1985). Self-organisation, selective adaptation, and its limits. In D.J. Depew and B.H. Weber (eds.), *Evolution at a Crossroads*. Cambridge, MA: MIT Press, pp. 169–207.

Keys, D.N., Lewis, D.L., Selegue, J.E., Pearson, B.J., Goodrich, L.V., Johnson, R.L., Gates, J., Scott, M.P., and Carroll, S.B. (1999). Recruitment of a *Hedgehog* regulatory circuit in butterfly eyespot evolution. *Science* 283, 532–4.

Kirschner, M. and Gerhart, J. (1998). Evolvability. *Proc Natl Acad Sci USA* 95, 8420–7.

Langlade, N.B., Feng, X., Dransfield, T., Copsey, L., Hanna, A.I., Thebaud, C., Bangham, A., Hudson, A., and Coen, E. (2005). Evolution through genetically controlled allometry space. *Proc Natl Acad Sci USA* 102, 10221–6.

Maynard Smith, J., Burian, R., Kaufman, S., Alberch, P., Campbell, J., Goodwin, B., Lande, R., Raup, D., and Wolpert, L. (1985). Developmental constraints and evolution. *Q Rev Biol* 60, 265–87.

McClain, C.R. (2005). Bathymetric patterns of morphological disparity in deep-sea gastropods from the western North Atlantic Basin. *Evolution* 59, 1492–9.

McClain, C.R., Johnson, N.A., and Rex, M.A. (2004). Morphological disparity as a biodiversity metric in lower bathyal and abyssal gastropod assemblages. *Evolution* 58, 338–48.

McGhee, G.R. (1999). *Theoretical Morphology*. New York: Columbia University Press.

(2007). *The Geometry of Evolution: Adaptive Landscapes and Theoretical Morphospaces*. Cambridge: Cambridge University Press.
Monteiro, A., Brakefield, P.M., and French, V. (1994). The evolutionary genetics and developmental basis of wing pattern variation in the butterfly *Bicyclus anynana*. *Evolution* 48, 1147–57.
(1997). Butterfly eyespots: the genetics and development of the color rings. *Evolution* 51, 1207–16.
Nijhout, H.F. (1991). *The Development and Evolution of Butterfly Wing Patterns*. Washington, DC: Smithsonian Institute Press.
Raup, D.M. (1966). Geometric analysis of shell coiling: general problems. *J Paleontol* 40, 1178–90.
(1967). Geometric analysis of shell coiling: coiling in ammonoids. *J Paleontol* 41, 43–65.
Reed, R.D., and Serfas, M.S. (2004). Butterfly wing pattern evolution is associated with changes in a Notch/Distal-less temporal pattern formation process. *Curr Biol* 14, 1159–66.
Ridley, M. (2003). *Evolution* (third edition). Oxford: Blackwell Publishing.
Robertson, K.A. and Monteiro, A. (2005). Female *Bicyclus anynana* butterflies choose males on the basis of their dorsal UV-reflective eyespot pupils. *Proc R Soc Lond B* 272, 1541–6.
Saenko, S.V., French, V., Brakefield, P.M., and Beldade, P. (2008). Conserved developmental processes and the formation of evolutionary novelties: examples from butterfly wings. *Philos Trans R Soc Lond B Biol Sci* 363, 1549–55.
Schluter, D. (1996). Adaptive radiation along genetic lines of least resistance. *Evolution* 50, 1766–74.
Stevens, M. (2005). The role of eyespots as anti-predator mechanisms, principally demonstrated in the Lepidoptera. *Biological Reviews* 80, 573–88.
Stone, J.R. (1996). Computer-simulated shell size and shape variation in the Caribbean land snail genus *Cerion*: a test of geometrical constraints. *Evolution* 50, 341–7.
Thompson, J.N. (2005). *The Geographic Mosaic of Coevolution*. University of Chicago Press.
Via, S. and Lande, R. (1985). Genotype–environment interaction and the evolution of phenotypic plasticity. *Evolution* 39, 505–22.
Wagner, A. (2005). *Robustness and Evolvability in Living Systems*. New Jersey: Princeton University Press.
Wagner, G.P. and Altenberg, L. (1996). Complex adaptations and the evolution of evolvability. *Evolution* 50, 967–76.
Wagner, P.J. (1995). Testing evolutionary constraint hypotheses with early Paleozoic gastropods. *Paleobiology* 21, 248–72.
Wagner, P.J. and Erwin, D.H. (2006). Patterns of convergence in general shell form among Paleozoic gastropods. *Paleobiology* 32, 316–37.
Weatherbee, S.D., Nijhout, H.F., Halder, G., Galant, R., Selegue, J., and Carroll, S. B. (1999). Ultrabithorax function in butterfly wings and the evolution of insect wing patterns. *Current Biology* 9, 109–15.

6

Innovation and diversity in functional morphology

PETER C. WAINWRIGHT

INTRODUCTION

Form and function are so intricately intertwined that it is tempting to view them as nothing more than different viewpoints on the phenotype – one reflecting the other like a mirror. And it is true that some organismal function can be revealed right there in the anatomy. But life is more complex than this, and while form does play a huge role in how structures are used and how they perform, it is not the only factor involved, and, in fact, the mapping of form to function can be surprisingly complex, with considerable subtlety. Emergent mechanical properties of structural systems can often be inferred or directly measured from form, but when it comes to actual function – how the structure is used by the organism – few systems are simple enough and so constrained that function can be completely inferred from form. Thus, while mechanical advantage of muscle acting across a joint can be calculated accurately from morphology, one could not predict the kinematics of locomotion in a quadrupedal vertebrate from limb morphology. The degrees of freedom in motion are too many and ultimately function must be measured in any but the simplest system before insight can be gained into how form influences function. Indeed, much of our intuition of form–function relationships is developed from studying the relationship empirically in one organism and then assuming certain commonalities in making predictions about unstudied organisms. This is not the same as being able to build predictions of function directly from form.

While it is not possible to precisely predict function from form in a complex system, it is true that the range of potential functions is set by the morphology. In this way, form determines a range of possible functions and the organism must select from this "potential function space" in shaping realized actions and uses of structures. Form determines potential for function, but organisms usually have choices about how they actually use a structure.

What are the implications of this organization for evolutionary changes in biomechanical and physiological systems? How do novelties impact the form–function relationship? They change the range of potential function and in doing so they may also open up new avenues to diversification of the system. My aim in this chapter is to consider innovations in the context of the form–function map. I offer a framework for this topic and hope to show that subtleties of the mapping of form to function can have a major impact on the dynamics of innovation in complex functional systems.

INNOVATION

Not all evolutionary novelties are equally potent. Some, such as the horns on the head of some chameleons, seem to have little effect on the ecological potential and success of the lineage in possession, while others, such as powered flight or endothermy, are major breakthroughs in functional design that drastically changed the ecology and evolution of the lineages in which they evolved. The history of life is characterized by the periodic introductions of novelties that seem to have had significant effects on subsequent ecological and evolutionary diversity: multicellularity, genome duplications, body segmentation, flowers, jaws, and so on. To the extent that innovations are an important causative agent in spurring bouts of morphological and ecological diversification, the study of innovation takes on great significance as we try to understand the uneven distribution of diversity across the tree of life. But how exactly do innovations influence diversity, and how do we go about testing hypotheses about their effects on macroevolution? These are the issues that I wish to address in this chapter.

From the outset, I wish to draw a distinction between species richness and other forms of diversity, such as morphological diversity, functional diversity, or ecological diversity. This chapter is about these latter forms of diversity, particularly morphological diversity. I will begin with a discussion of two major types of innovations that influence morphological diversity: those that change the potential morphology of the organism – the potential morphospace – and those that change the adaptive landscape that the organism is exposed to. In addition, I will illustrate that the nature of how morphology maps onto function can influence patterns of diversification in evolving lineages. Innovations in organismal design that affect this mapping have emerged as a new class of innovations that can impact phenotypic diversity. The science of doing comparative analyses of morphological and ecological diversity has lagged behind similar studies of lineage diversification rate, or species richness, and I review some recent methodological and conceptual progress on this front. I illustrate mechanisms of how innovations affect diversity and how one can test for the effects of innovations on diversity with a research example from my laboratory on the evolution of parrotfish feeding functional morphology. However, my aim is to make points that should generalize to any level of design, from molecular to whole organism.

I emphasize two categories of innovations in terms of how they impact diversity: those that directly influence the potential for morphological variation and those that allow the lineage to move into new regions of the adaptive landscape where new variants are favored. Innovations in the first category change the potential morphospace of the body plan possessed by the lineage. Innovations in the second category represent a breakthrough in organismal performance that allow the lineage to move into a novel region of the adaptive landscape where a variety of new adaptive peaks can be reached. These two classes of innovations mirror the fundamental distinction in morphospace biology between the morphospace of the theoretically possible (Hickman 1993; Raup 1966) and the adaptive landscape that is produced by mapping fitness into the morphospace (Arnold 2003; McGhee 1999; Wright 1932).

Growing the theoretical morphospace

The potential morphospace occupied by a body plan is determined by the number of independent parameters that are required to define the morphospace. In his classic work on the mollusk shell, Raup identified three parameters that generated a morphospace of all mollusk shells (Raup 1966). A novelty that increases the number of parameters required to describe the form increases the size of the potential morphospace and provides an opportunity for greater diversity. One conceptually simple novelty that results in this sort of increase in morphospace is a structural duplication or subdivision event that results in increases in the number of elements that make up the form, and thus increases the dimensionality of potential morphospace.

A common anatomical form of duplication is segmentation of the body, a phenomenon that typically involves replicated body regions each with the same basic plan. Segmentation illustrates the general ways in which duplication enhances diversity. Repeated body segments allow retention of a role in a primitive function in one or more segments, while other segments can become modified for novel functions. Arthropods offer a classic example of this pattern, in which the body is segmented into units that each possess axial structures and limb elements. Anterior body segments are modified for performance in sensory systems and jaws, and more posterior segments for locomotor specialization. In decapod shrimps, some body segments are modified for walking, while other segments are specialized for burst locomotion, which is used during escape from predators. In arthropods, the duplication of body segments with the same suite of anatomical elements permits an expansion of the potential morphospace, since each element in each segment becomes a new axis in morphospace. In practice, there have been relatively few case studies that explore the consequences of duplication or subdivision events for morphospace expansion and lineage diversity (Friel and Wainwright 1997; Schaefer and Lauder 1986, 1996).

Redundancy creates the potential for different body segments to perform different functions, such that the specialization of a segment for, say, burst locomotion, need not be constrained by a need for the same segment to maintain performance in a second function, such as walking behavior. This consequence of duplication events, known as functional decoupling, can enhance diversity because it more readily leads to body designs that exhibit higher overall performance capacities. Thus, for example, when walking and burst swimming are performed by different body regions, adaptation for higher performance is not constrained by the need for a single region to maintain both functions. Functional decoupling through redundancy in design is one of the most widespread and powerful ways in which novelties become innovations that lead to increased diversity. This mechanism is well known in molecular evolution (Burmester *et al.* 2006; Cao *et al.* 2006; Chung *et al.* 2006; Lynch 2003; Ohno 1970; Spady *et al.* 2005), developmental biology (Carroll 2001; Hughes and Friedman 2005), and organismal functional design (Emerson 1988; Friel and Wainwright 1997; Lauder 1990; Schaefer and Lauder 1996).

The idea that a single anatomical system may be required to perform multiple functions has been viewed as a major constraint on evolutionary diversification; and, conversely, the decoupling of such a constraint is seen as a major avenue to increasing diversity. While structural duplication is conceptually the simplest way to achieve this sort of decoupling, there are other routes. Gatesy and Middleton (1997) argued that the introduction of winged locomotion in birds released a constraint on the hindlimbs of theropods. Their proposal was that the theropod hindlimb had been morphologically constrained because this body region was the sole system used for locomotion. With the origin of forelimb-powered locomotion (flight), they argued that the bird hindlimb was free to become modified for a greater diversity of locomotor specializations. They showed evidence for increased diversity of the hindlimb in birds as compared to theropods (Gatesy and Middleton 1997).

Importance of the form–function map

In recent years there has been mounting evidence of intrinsic design features that can relieve functional constraints on anatomical systems in the absence of duplication events. One general mechanism is seen in what is referred to as many-to-one mapping of morphology to function (Alfaro *et al.* 2004, 2005; Hulsey and Wainwright 2002; Wainwright *et al.* 2005). The idea here is that one inherent feature of a complex functional system is that multiple morphologies can have the same functional property. This occurs when the functional property depends on three or more underlying parameters and is contrasted with simple systems that are determined by one or two parameters. Take the simple case of the mechanical advantage of a muscle acting across a joint. Mechanical advantage is the ratio of two distances: an input lever length and an output lever length. Once scale is removed, there is only a single combination of input lever length and output lever length that results in any particular value of mechanical advantage. This system exhibits one-to-one mapping. Now consider instead the force exerted at the end of the output lever. This is a function of the mechanical advantage of the lever and a third parameter, the input force exerted by the muscle acting on the lever. Output force is equal to input force times the ratio of input lever to output lever. Here there are numerous combinations of the three parameters that all give the same value of output force. For example, mechanical advantage of 0.5 (input lever length/output lever length=0.5/1) and input force of 8 give the same output force as a mechanical advantage of 1.0 (input lever/output lever=1/1) and an input force of 4. In this case, there is a many-to-one mapping of musculoskeletal design to force output of the system.

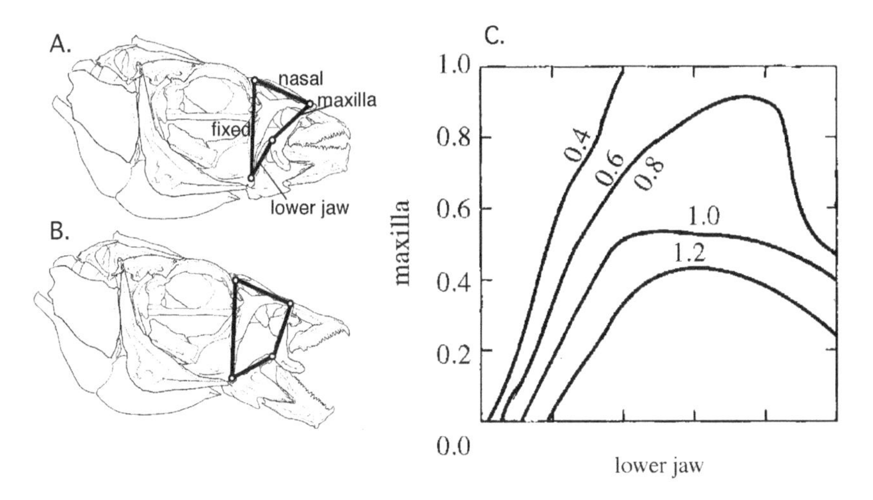

Figure 6.1 A & B. Illustration of the four-bar linkage lever system that governs movement of the upper and lower jaws of labrid fishes. The transmission coefficient of interest in this four-bar is called "KT" and is equal to the number of degrees of output rotation in the maxilla bone divided by the number of input degrees of rotation of the lower jaw. C. A slice through the morphospace of all possible four-bar shapes, showing the isoclines of identical values of KT. Many four-bar shapes have the same mechanical property, thus illustrating the property of many-to-one mapping of morphology onto the functional property.

Many-to-one mapping has been described in the mechanics of four-bar linkages, a type of lever system that involves four stiff elements connected together in a loop (fig. 6.1; Hulsey and Wainwright 2002). Like simple levers, four-bar linkages transmit force and motion from one element of the system to other elements. A four-bar linkage is found in the jaws of some fishes that transmits movement and force from the lower jaw to the upper jaw (fig. 6.1). The mechanical advantage of this lever system exhibits many-to-one mapping as many four-bar shapes have the same mechanical advantage (fig. 6.1C).

The implications of many-to-one mapping of form to function for diversity are at least twofold. First, because many forms can have the same functional property, there are surfaces in morphospace that define mechanically neutral variation. These regions of neutral change in morphology offer opportunities for lineages to explore novel regions in morphospace without paying fitness costs. A second consequence involves the observation that most body parts serve multiple functions. This means that an anatomical region may have multiple functional properties that are all potentially exposed to natural selection. Many-to-one mapping and the resulting regions of neutral morphological variation with respect to one functional property allow flexibility in design so that the structure can become modified for other functions while maintaining their performance on the original function.

Breakthroughs that change the adaptive landscape

While changes in body plan can alter the range of potential morphologies, it is also clear that some novelties have sufficiently radical consequences for the performance capacity of the organism that they effectively open up whole new possible ways of life and can lead to subsequent diversification. Sometimes referred to as "major innovations" or "key innovations," such an innovation is a breakthrough in design that moves the lineage into a new region of the adaptive landscape. There are many compelling examples. With the origin of powered flight in birds, new feeding strategies, life-history patterns, and opportunities for novel habitat use were made possible, and the resulting radiation appears to owe much of its diversity to this design breakthrough. Similarly, the origin of jaws in vertebrates was followed by a successful radiation of highly predatory organisms. Note that in these examples, the innovations involve a significant enhancement in some aspect of the behavioral performance capacity of the organisms that opened up the exploitation of novel resources, such as new food types, new habitats, or new life-history patterns.

Both increases in the potential morphospace and breakthroughs in design only open up the potential for subsequent radiations; they do not make such radiations inevitable. There is an important role for stochastic processes and the appropriate ecological conditions of disruptive selection to realize the potential change in diversity. In this sense, the innovations only set the stage for changes in diversity; they do not, by themselves, cause the change. Neither do changes in morphological diversification need to be tied to changes in speciation rate, extinction rate, or net diversification rate (Ricklefs 2004, 2006).

COMPARATIVE ANALYSIS OF DIVERSITY

The discussion above suggests a research program in the history and biology of innovations. There are many questions that one might like to ask about a putative innovation and its consequences for diversity. Was the innovation followed by an increase in the diversity of the functional systems affected by the breakthrough, or of other functional systems? Does the innovation result in a qualitative shift in some aspect of performance capacity or resource use? Given that the innovation is associated with an increase in diversity, what is the tempo of that change? Is there a period of relative stasis before the diversity is accelerated, or does the change happen coincident with the innovation? These questions are inherently historical, and all of them can be addressed with the use of a phylogeny of the group and its relatives as a basis for comparisons. Methods and concepts for conducting phylogenetically correct comparisons of morphological, functional, and ecological diversity (Garland *et al.* 1992; O'Meara *et al.* 2006) between lineages have lagged behind the development of methods for comparative studies of species diversity (Sanderson and Donoghue 1994; Slowinski and Guyer 1994). Important insights about the role of phylogenetic history in species diversity have resulted in the emergence in recent years of a focus on lineage diversification rate. But, just as diversification rate should be recognized as the phylogenetically corrected metric of species richness in a clade, similarly, rate of morphological evolution provides a phylogenetically corrected metric of trait evolution (Martins 1994; O'Meara *et al.* 2006). Below, I review an approach for comparing rates of morphological evolution – that is, morphological diversity – between two lineages. This is followed by a test-case example involving the feeding biology of parrotfish.

Morphological diversity

There are several widely used metrics of morphological diversity, but variance and range are the most widely used (reviewed by Foote 1997). Range is of interest because it reflects information about the furthest regions of morphospace that have been reached by members of the group in question (Pie and Weitz 2005). It may be useful in addressing questions about what regions of morphospace have been occupied by a group and which have not (Stebbins 1951; Van Valkenburgh 1988). The multivariate measure of range is usually some version of an N-dimensional minimum polygon that encloses all individuals in the group. While range is of particular interest in some case studies, the statistical properties of a range make doing careful quantitative comparisons between groups awkward. Also, in a Gaussian distribution, range scales with sample size, further complicating comparisons.

Variance of traits is the most widely used characterization of morphological diversity (Foote 1997; McClain 2005; Roy and Foote 1997). Variance captures the dispersion of members of the group in morphospace, is not so susceptible to the effects of a few outliers, and does not scale with sample size, so the metric is versatile and amenable to statistical testing. As we shall see, variance also relates directly to most models of character evolution, such as Brownian motion, so that the connection between the model of evolution and variance among evolving lineages is strong (O'Meara *et al.* 2006).

It is intuitive that morphological diversity among species is affected by their phylogenetic history. After all, species usually most resemble their closest relatives. But exactly how do we expect phylogeny to relate to morphological diversity, and how can we use this knowledge in framing comparative tests of morphological diversity? To get at these issues, we first need a model of trait evolution. Perhaps the most straightforward model used is Brownian motion (Martins and Hansen 1997) – that is, the model used in calculating independent contrasts (Felsenstein 1985) and the estimation of ancestral states (Schluter *et al.* 1997). Under this model of evolution, the potential for change in the trait occurs at some designated time interval, with the magnitude of the change being drawn from a normal distribution with mean of zero and some variance. The variance of this distribution of potential trait change is referred to as the Brownian rate parameter. The expected variance of the trait among lineages in a phylogeny is equal to the number of opportunities for trait change (proportional to time in the Brownian model) times this variance in the distribution of potential trait changes, or the Brownian rate parameter. The larger the rate parameter, the greater the variance among lineages we expect to see. Thus, diversity among members of a lineage emerges as a function of the time in the phylogeny, the amount of shared history that lineages share, and the rate of evolution of the trait. A key insight that emerges from this is that a phylogenetically corrected comparison of morphological diversity between two clades involves removing the confounding effects of time and shared history, and comparing the rate of evolution of the traits of interest. When the Brownian motion model of trait evolution is used, this means comparing the estimates of the Brownian rate parameter that are provided by distribution of trait values among the terminals on the phylogeny.

As a simple illustration of this effect of time, consider two monophyletic groups of birds: one that shows considerable variation among species in bill morphology, and an other that shows minimal differences among species (fig. 6.2). In each group, bill morphology has been diversifying since the time of the most recent common ancestor (MRCA). If the age of the MRCA of the diverse group is considerably higher than the age of the MRCA in the low-diversity group (fig. 6.2A), then time may be a trivial explanation for the difference in diversity between groups. But if the MRCAs are of similar age (fig. 6.2B), or the low-diversity group is actually older, then we would infer that the rate of evolution of bill morphology has been higher in the diverse group.

This framework is formalized in recently developed software that accepts as input a phylogeny with branch lengths in time, or relative time, and trait values for the tips of the phylogeny (Collar *et al.* 2005; O'Meara *et al.* 2006). The program then estimates the Brownian rate parameter and allows one to compare it between two clades, or between a clade and its paraphyletic outgroup. This program is well designed for testing hypotheses of the effects of specific innovations, or synapomorphies, on morphological diversity. In these tests, it is actually the Brownian rate parameter, or the rate of trait evolution, that is compared between groups, thus removing the confounding effects of time and shared history.

Case study: innovation and diversity in the feeding mechanism of parrotfish

Parrotfish, or Scaridae, are an icon for the biodiversity found on coral reefs. Members of this monophyletic group of about a hundred species are found on coral reefs and reef-associated habitats around the world, where they are among the most numerically dominant and ecologically important fishes (Bellwood 1994; Brock 1979). Normally thought of as herbivores, parrotfishes are omnivores that feed by removing algae and small benthic invertebrates that grow on the surface of the reef (Choat *et al.* 2004; Randall 1967). Many species have beak-like jaws that are used in scraping or gouging the surface of growth-covered coral skeletons or directly clipping large algae. The grazing activities of the parrotfish have been shown to be crucial to the success of reef-building corals in gaining a foothold in benthic communities, because many algae are competitively dominant to coral if allowed to grow unchecked (Lewis and Wainwright 1985). Reefs which have lost their parrotfish become overgrown with algae and also lose their coral. The Scaridae are a monophyletic group (Bellwood 1994), nested within the larger Labridae (Westneat and Alfaro 2005).

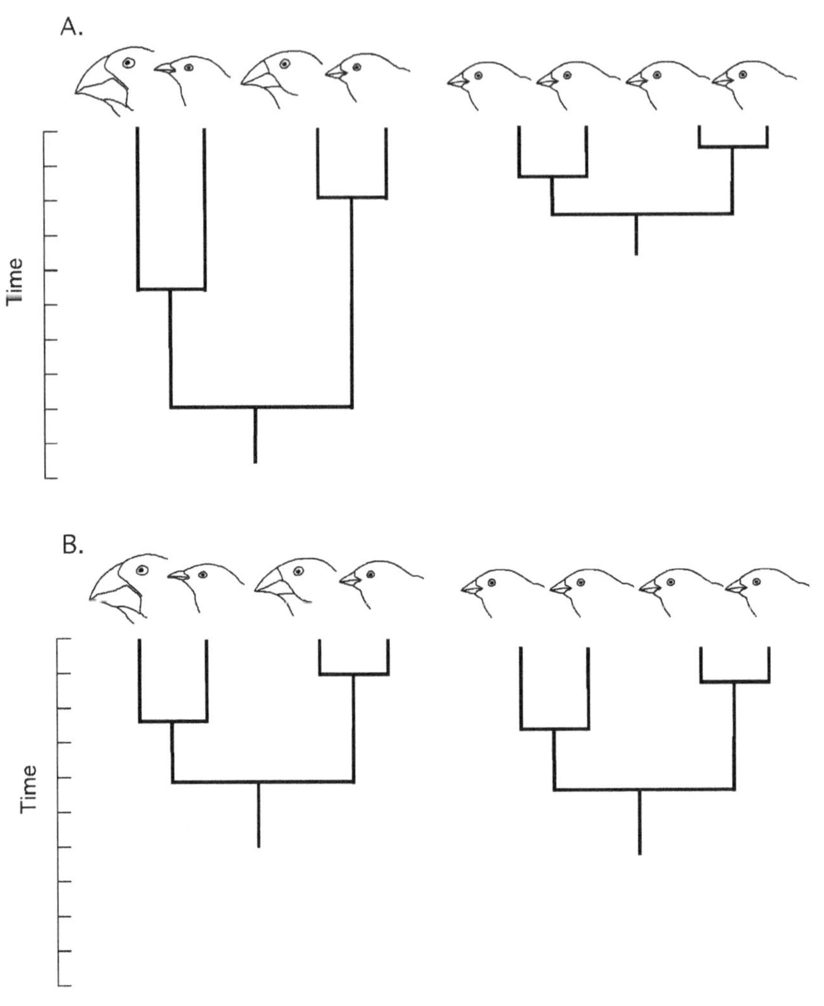

Figure 6.2 Diagrams illustrating the importance of time in the interpretation of differences in morphological diversity between two monophyletic groups of birds. A. The high-diversity group is much older than the low-diversity group. In this case, the difference in diversity between groups may be due to differences in the amount of time the two groups have had to diversify. B. Here, the age of the two groups and the total time in the phylogenies are the same, suggesting that the rate of bill evolution in the high-diversity group would have been higher than in the low-diversity group. The approach described in this chapter is designed to separate the effects of time and rate of evolution on the observed diversity in a group of terminal taxa. Figure is modified from Wainwright 2007.

An apparent key to the trophic success of parrotfishes is their ability, unique among fishes, to pulverize and grind the combination of algae, encrusting invertebrates and coral rock that they ingest. A pulverized slurry of ground coral skeleton, algae, and inverterbates is mixed with secretions from an active mucous gland in the pharynx and passed into the digestive tracts, where digestion occurs (Gobalet 1980, 1989). The ground coral skeletons pass through the gut undigested and are excreted back onto the reef (Bellwood 1996). The constancy of this activity, and the high biomass of parrotfishes, results in them being the primary producers of sand on many coral reefs around the world (Bellwood and Choat 1990; Bellwood *et al.* 2003).

The capacity of parrotfishes to pulverize and grind their food is due to a suite of novelties in their pharyngeal jaw apparatus (fig. 6.3) that allows them to have both a very strong pharyngeal bite, and the capacity to cyclicly protract and retract the upper pharyngeal jaw while occluding the jaws (Gobalet 1980, 1989; Wainwright 2005). This simultaneous anterior-posterior motion of the upper jaw, while the lower jaw is firmly adducted against the upper jaw, is used to grind and pulverize the scrapings that the parrotfish takes from the reef. The scarid pharyngeal jaw possesses several major muscular and skeletal modifications that contribute to this functionality (Bellwood 1994; Gobalet 1989). These modifications of the pharyngeal jaw, and the ability to use the pharyngeal jaw to grind and pulverize prey, are synapomorphies for the Scaridae. Neither the modifications nor the ability to grind up their prey so extensively are found in other labrid fishes.

These innovations in the pharyngeal jaw represent a breakthrough in design of the parrotfish feeding mechanism that allows them access to novel food resources not exploited by other labrid fishes. I would like to explore the hypothesis that by making algae-grazing a viable trophic strategy, this breakthrough led to a burst in morphological and functional diversification of the scarid oral jaws that are used in procuring food. With the novel ability to feed successfully by grinding up the communities of algae and small invertebrates that live on the surface of corals, have parrotfishes experienced an adaptive radiation involving increased diversity of the oral jaws for a range of prey-capture specializations?

Among parrotfishes there is considerable variation in the details of how their benthic prey are removed from the substrate. Some taxa are best described as grazers, because they lightly scrape the surface of rock and sand to remove algae, mats of bacteria, or detritus. Other species,

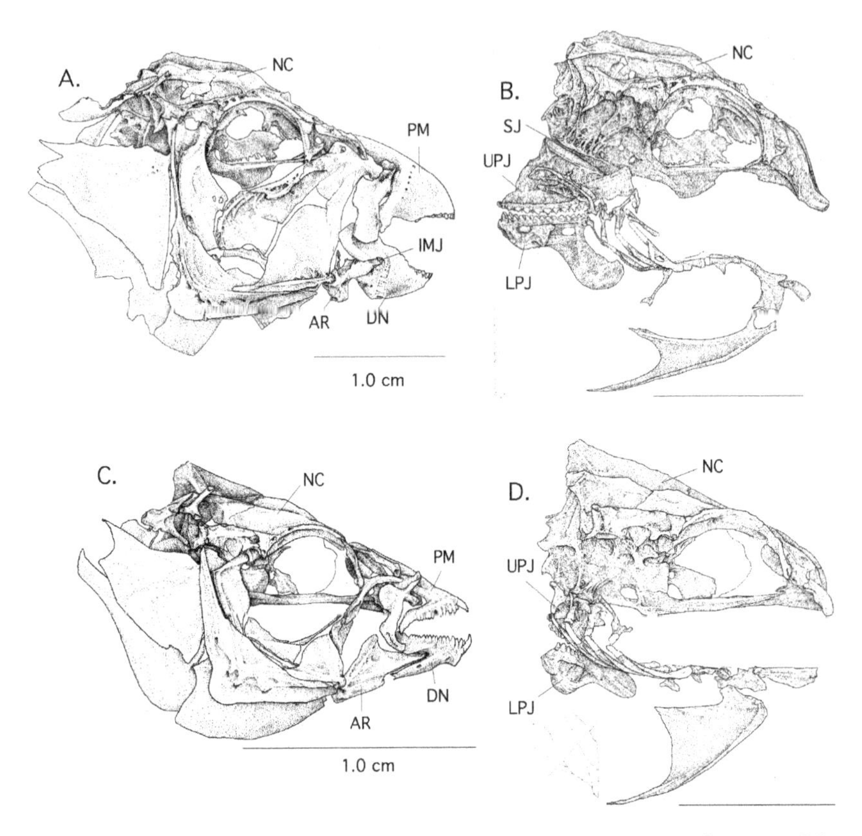

Figure 6.3 Diagrams of the skull and pharyngeal jaw apparatus of a parrotfish (A & B) and a non-scarid labrid (C & D). A. Skull of the parrotfish *Chlorurus sordidus*, demonstrating the beak-like jaw formed by fusion of teeth, and the intramandibular joint that enhances the vertical gape during mouth-opening and allows the parrotfish to fine-tune the orientation of the beak through the closing cycle. B. Pharyngeal apparatus of *C. sordidus*, located deep to the structures in A. The parrotfish's pharyngeal jaw has an expanded toothed surface and a greatly enlarged sliding joint between the upper jaw and the underside of the neurocranium. C. Skull of the non-scarid labrid *Cheilinus oxycephalus* to illustrate a typical labrid. This fish is a suction-feeder and lacks an intramandibular joint in the lower jaw. D. Pharyngeal apparatus of *C. oxycephalus*, showing the typical non-scarid condition in labrids. The jaws are smaller than in the parrotfish, and the articulation between upper jaw and neurocranium is a small raised surface on the underside of the skull that rests in a depression on the upper jaw, forming a pivoting joint. Figure and data are modified from Wainwright, Bellwood, Collar, and Alfaro (unpublished manuscript). Abbreviations: AR, articular; DN, dentary; IMJ, intramandibular joint; LPJ, lower pharyngeal jaw; NC, neurocranium; PM, premaxilla; SJ, sliding joint; UPJ, upper pharyngeal jaw.

referred to as gougers, dig their beak-like jaws deep into the reef and excavate big pieces of dead coral rock, which are ground up for the algae and associated fauna of boring and encrusting invertebrates. Some species feed on the epiphytes that grow on seagrass blades, while others feed extensively on live coral. This diversity of feeding methods is associated with variation in several features of the oral jaws. Several taxa possess an intramandibular joint in the lower jaw that may enhance force delivery of the jaw-closing adductor mandibulae muscle (fig. 6.3A). As mentioned above, some taxa have a beak-like jaw formed by fusion of the teeth, while other taxa have more typical labrid teeth.

To test the hypothesis that morphological diversity is greater in parrotfishes than in other labrids, we estimated and compared between parrotfish and non-parrotfish labrids the rates of evolution of seven morphological traits from the oral jaws (Wainwright *et al.* unpublished manuscript). Morphological measurements were made on at least three specimens per species, for thirteen parrotfish species and fifty species of non-parrotfish labrids (fig. 6.4). We measured the mechanical advantage of jaw-opening and jaw-closing movement, the diameter of the open mouth, the amount of jaw protrusion, the mass of the adductor mandibulae (the jaw-closing muscle), the mass of the sternohyoideus muscle (used to expand the buccal cavity during suction-feeding), and the transmission coefficient of the oral-jaw four-bar linkage (Wainwright *et al.* 2004). Average values of the morphological traits per species were used in the analyses. A phylogeny of the sixty-three species was generated by combining DNA sequence data used in a study of the Labridae (Westneat and Alfaro 2005) with data used in a study on parrotfishes (Streelman *et al.* 2002). The resulting phylogeny (fig. 6.4) was made ultrametric using the program r8s (Sanderson 2003), and further analyses used this tree topology, with branch lengths reflecting relative time. The phylogeny indicates that parrotfishes are a monophyletic group nested within the Labridae, and that this latter group is therefore paraphyletic unless it includes the scarids. The morphological characters were log-transformed, and body size was removed by calculating residuals from regressions with the log of fish standard length. Rates of evolution of the seven morphological variables were calculated with the program Brownie (O'Meara *et al.* 2006), which allowed us to perform likelihood ratio tests of the difference between groups in rate, by forming a ratio of the likelihood score for a one-parameter model (single-rate parameter fit for the entire tree) with the likelihood for a two-parameter model (separate rates for the non-parrotfish

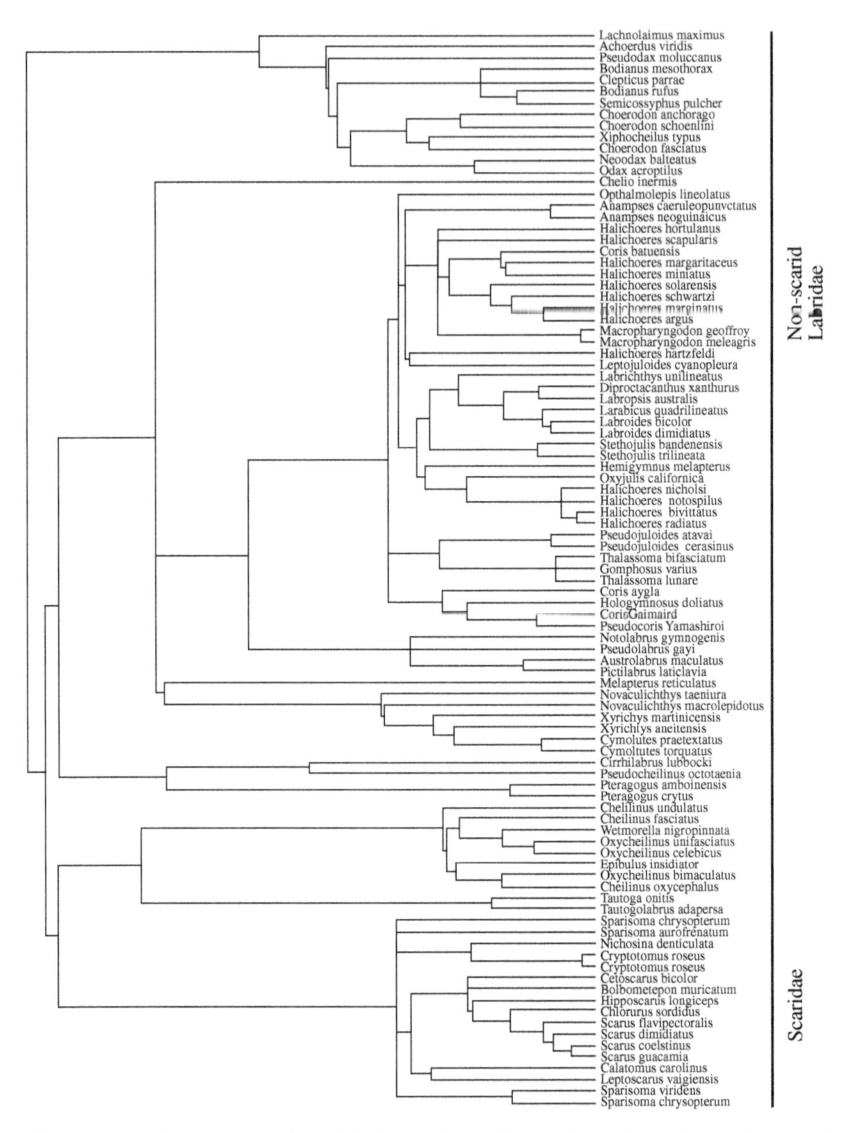

Figure 6.4 Chronogram of the labrids and scarids used in this analysis of rates of evolution of oral-jaw characters. Scaridae are nested within the Labridae, making the latter group paraphyletic. In this analysis, rates of evolution of jaw traits were estimated in the Scaridae and compared to the paraphyletic non-scarid Labridae. Figure is modified from Wainwright, Bellwood, Collar, and Alfaro (unpublished manuscript).

labrids and the scarids). Casual inspection of the ultrametric phylogram reveals that the age of the most recent common ancestor of the non-parrotfish labrids is older than the most recent common ancestor of the parrotfishes (fig. 6.4).

An initial comparison of oral-jaw morphological diversity in parrotfishes with that in the non-parrotfish labrids (fish known as wrasses) reveals a pattern in which some traits trend toward higher variance among species in parrotfish (jaw-closing mechanical advantage, jaw protrusion), while others have slightly higher standing variance in wrasses (maxillary transmission coefficient, mouth diameter, adductor mandibulae muscle mass, sternohyoideus muscle mass). Since the age of labrids is quite a bit older than scarids (fig. 6.4), time clearly confounds these comparisons. The likelihood ratio test indicates that there is a significant difference between the groups in rate of evolution of two traits – the jaw-closing mechanical advantage and upper-jaw protrusion (fig. 6.5). Both variables have a higher rate of evolution in scarids than in the non-parrotfish labrids. Jaw-closing mechanical advantage evolved about seven times faster in scarids, and jaw protrusion evolved about six times faster in scarids. Jaw-opening mechanical advantage evolved about twice as fast in scarids, although this difference was not significant ($P=0.09$).

These results indicate that the rate of evolution of two of seven oral-jaw characters related to feeding mechanics have evolved faster in scarids than in other labrids, while a third trait shows a marginally higher rate of evolution in scarids. Thus, following the origin of the modified parrotfish pharyngeal jaw apparatus, the mechanics of the oral jaw have undergone a more rapid functional diversification than has been seen across other labrids. This difference between scarids and non-scarid labrids in the rate of evolution of oral-jaw traits is particularly noteworthy given that wrasses are widely known for being perhaps the most trophically diverse group of fishes that occurs on coral reefs. Various species of wrasses feed on other fishes, decapod shrimps and crabs, copepods, polychaete worms, mollusks, foraminifera, coral mucous, ectoparasites of other fishes, and hermit crabs (Bellwood *et al.* 2006; Randall 1967; Wainwright *et al.* 2004). Nearly every trophic niche occupied by reef fishes can be seen among wrasses, and extensive morphological diversity is seen that mirrors this diversity. Yet once we account for the effects of time and shared history, parrotfishes have greater diversity in the mechanics of their oral jaws than wrasses.

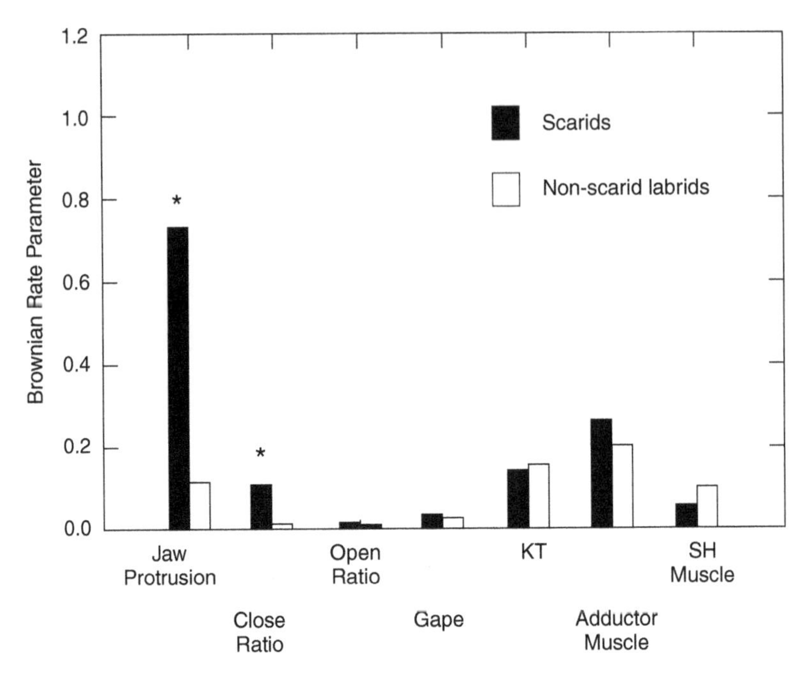

Figure 6.5 Bar diagram illustrating the estimated Brownian rate parameters for the seven oral-jaw traits included in this study. Jaw protrusion and the jaw-closing lever ratio showed significantly higher rates of evolution in parrotfishes than in the non-scarid labrids. The asterisk indicates that the likelihood ratio test was significant at $P < 0.0001$. Rate of evolution of the adductor mandibulae was also marginally higher in the Scaridae ($P = 0.09$). These results indicate that the parrotfish oral jaws have experienced higher rates of evolution than the same features in the non-scarid labrids, supporting the prediction that the modified pharyngeal jaw of the parrotfish that confers the novel ability to feed on algae and small encrusting invertebrates has resulted in an adaptive radiation of the parrotfish oral jaws. Figure is modified from Wainwright, Bellwood, Collar, and Alfaro (unpublished manuscript).

CONCLUSIONS

Innovations can have drastic effects on the diversity of evolving lineages, whether they change the nature of the morphospace that describes the organism, or change the nature of diversifying selection that a lineage is exposed to. But in order to test hypotheses about the effects of innovations on morphological diversity, the confounding effects of time and amount of shared history must be separated from changes in the rate of morphological evolution. This can be accomplished using the new computer program Brownie, which models trait evolution under Brownian motion, and allows the user to estimate and compare the rate of evolution of characters between two groups in a given phylogeny. Just as net diversification rate controls for time and tree structure in studies of species richness, measures of the rate of trait evolution control for the effects of time and shared history in comparative studies of morphological, functional, or ecological diversity. These recently developed methods offer a powerful new tool in the study of innovations and provide one approach to distinguishing adaptive radiation from background rates of morphological evolution.

REFERENCES

Alfaro, M.E., Bolnick, D.I., and Wainwright, P.C. (2004). Evolutionary dynamics of complex biomechanical systems: an example using the four-bar mechanism. *Evolution* 58, 495–503.

(2005). Evolutionary consequences of many-to-one mapping of jaw morphology to mechanics in labrid fishes. *American Naturalist* 165, E140–E154.

Arnold, S.J. (2003). Performance surfaces and adaptive landscapes. *Integrative and Comparative Biology* 43, 367–75.

Bellwood, D.R. (1994). A phylogenetic study of the parrotfishes family Scaridae (Pisces: Labroidei), with a revision of genera. *Records of the Australian Museum Supplement* 20, 1–86.

(1996). Production and reworking of sediment by parrotfishes (family Scaridae) on the Great Barrier Reef, Australia. *Marine Biology* 125, 795–800.

Bellwood, D.R. and Choat, R. (1990). A functional analysis of grazing in parrotfishes (family Scaridae): the ecological implications. *Environmental Biology of Fishes* 28, 189–214.

Bellwood, D.R., Hoey, A.S., and Choat, J.H. (2003). Limited functional redundancy in high diversity systems: resilience and ecosystem function on coral reefs. *Ecology Letters* 6, 281–5.

Bellwood, D.R., Wainwright, P.C., Fulton, C.J., and Hoey, S.A. (2006). Functional versatility supports coral reef biodiversity. *Proceedings of the Royal Society of London, Series B* 273, 101–7.

Brock, R.E. (1979). An experimental study of the effects of grazing by parrotfishes and role of refuges in benthic community structure. *Marine Biology* 51, 381–8.

Burmester, T., Storf, J., Hasenjager, A., Klawitter, S., and Hankeln, T. (2006). The hemoglobin genes of Drosophila. *FEBS Journal* 273, 468–80.

Cao, Z.J., Luo, F., Wu, Y.L., Mao, X., and Li, W.X. (2006). Genetic mechanisms of scorpion venom peptide diversification. *Toxicon* 47, 348–55.

Carroll, S.B. (2001). Chance and necessity: the evolution of morphological complexity and diversity. *Nature* 409, 1102–9.

Choat, J.H., Robbins, W.D., and Clements, K.D. (2004). The trophic status of herbivorous fishes on coral reefs – II. Food processing modes and trophodynamics. *Marine Biology* 145, 445–54.

Chung, W.Y., Albert, R., Albert, I., Nekrutenko, A., and Makova, K.D. (2006). Rapid and asymmetric divergence of duplicate genes in the human gene coexpression network. *BMC Bioinformatics* 7, 46.

Collar, D.C., Near, T.J., and Wainwright, P.C. (2005). Comparative analysis of morphological diversity: trophic evolution in centrarchid fishes. *Evolution* 59, 1783–94.

Emerson, S. (1988). Testing for historical patterns of change: a case study with frog pectoral girdles. *Paleobiology* 14, 174–86.

Felsenstein, J. (1985). Phylogenies and the comparative method. *American Naturalist* 125, 1–15.

Foote, M. (1997). The evolution of morphological diversity. *Annual Review of Ecology and Systematics* 28, 129–52.

Friel, J.P. and Wainwright, P.C. (1997). A model system of structural duplication: homologies of adductor mandibulae muscles in tetraodontiform fishes. *Systematic Biology* 46, 441–63.

Garland, T.J., Harvey, P. H., and Ives, A.R. (1992). Procedures for the analysis of comparative data using phylogenetically independent contrasts. *Systematic Biology* 41, 18–32.

Gatesy, S.M. and Middleton, K.M. (1997). Bipedalism, flight, and the evolution of theropod locomotor diversity. *Journal of Vertebrate Paleontology* 17, 308–29.

Gobalet, K.W. (1980). *Functional morphology of the head of parrotfishes of the genus Scarus*. Ph.D. thesis, University of California, Davis.

(1989). Morphology of the parrotfish pharyngeal jaw apparatus. *American Zoologist* 29, 319–31.

Hickman, C.S. (1993). Theoretical design space: a new program for the analysis of structural diversity. In A. Seilacher and K. Chinzei (eds.), *Progress in Constructional Morphology: Neues Jahrbuch für Geologie und Paläontologie – Abhandlungen*, pp.169–82.

Hughes, A.L. and Friedman, R. (2005). Gene duplication and the properties of biological networks. *Journal of Molecular Evolution* 61, 758–64.

Hulsey, C.D. and Wainwright, P.C. (2002). Projecting mechanics into morphospace: disparity in the feeding system of labrid fishes. *Proceedings of the Royal Society of London. Biological Sciences Series B* 269, 317–26.

Lauder, G.V. (1990). Functional morphology and systematics: studying functional patterns in an historical context. *Annual Review of Ecology and Systematics* 21, 317–40.

Lewis, S.M. and Wainwright, P.C. (1985). Herbivore abundance and grazing intensity on a Caribbean coral reef. *Journal of Experimental Marine Biology and Ecology* 87, 215–28.

Lynch, M. (2003). The origins of genome complexity. *Science* 302, 1401.

Martins, E.P. (1994). Estimating the rate of phenotypic evolution from comparative data. *American Naturalist* 144, 193–209.

Martins, E.P. and Hansen, T.F. (1997). Phylogenies and the comparative method: a general approach to incorporating phylogenetic information into the analysis of interspecific data. *American Naturalist* 149, 646–67.
McClain, C.R. (2005). Bathymetric patterns of morphological disparity in deep-sea gastropods from the western North Atlantic Basin. *Evolution* 59, 1492–9.
McGhee, G.R. (1999). *Theoretical Morphology: The Concept and its Applications (Perspectives in Paleobiology and Earth History)*. New York: Columbia University Press.
Ohno, S. (1970). *Evolution by Gene Duplication*. New York: Springer.
O'Meara, B., Ane, C.M., Sanderson, M.J., and Wainwright, P.C.. (2006). Testing for different rates of continuous trait evolution using likelihood. *Evolution* 60, 922–33.
Pie, M.R. and Weitz, J.S. (2005). A null model of morphospace occupation. *American Naturalist* 166, E1–E13.
Randall, J.E. (1967). Food habits of reef fishes of the West Indies. *Studies in Tropical Oceanography* 5, 655–847.
Raup, D.M. (1966). Geometric analysis of shell coiling: general problems. *Journal of Paleontology* 40, 1178–90.
Ricklefs, R.E. (2004). Cladogenesis and morphological diversification in passerine birds. *Nature* 430, 338–41.
(2006). Time, species, and the generation of trait variance in clades. *Systematic Biology* 55, 151–9.
Roy, K. and Foote, M. (1997). Morphological approaches to measuring biodiversity. *Trends in Ecology & Evolution* 12, 277–81.
Sanderson, M.J. (2003). r8s: inferring absolute rates of molecular evolution and divergence times in the absence of a molecular clock. *Bioinformatics* 19, 301–2.
Sanderson, M.J. and Donoghue, M.J. (1994). Shifts in diversification rate with the origin of angiosperms. *Science* 264, 1590–3.
Schaefer, S.A. and Lauder, G.V. (1986). Historical transformation of functional design: evolutionary morphology of the feeding mechanisms of loricariod catfishes. *Systematic Zoology* 35, 489–508.
(1996). Testing historical hypotheses of morphological change: biomechanical decoupling in loricarioid catfishes. *Evolution* 50, 1661–75.
Schluter, D., Price, T., Mooers, A.O., and Ludwig, D. (1997). Likelihood of ancestor states in adaptive radiation. *Evolution* 51, 1699–711.
Slowinski, J.B. and Guyer, C. (1994). Testing whether certain traits have caused amplified diversification: an improved method based on a model of random speciation and extinction. *American Naturalist* 142, 1019–24.
Spady, T.C., Seehausen, O., Loew, E.R., Jordan, R.C. Kocher, T.D., and Carleton, K.L. (2005). Adaptive molecular evolution in the opsin genes of rapidly speciating cichlid species. *Molecular Biology and Evolution* 22, 1412–22.
Stebbins, G.L. (1951). Natural selection and the differentiation of angiosperm families. *Evolution* 5, 299–324.
Streelman, J.T., Alfaro, M., Westneat, M.W., Bellwood, D.R., and Karl, S.A. (2002). Evolutionary history of the parrotfishes: biogeography, ecomorphology, and comparative diversity. *Evolution* 56, 961–71.

Van Valkenburgh, B. (1988). Trophic diversity in past and present guilds of large predatory mammals. *Paleobiology* 14, 156–73.

Wainwright, P.C. (2005). Functional morphology of the pharyngeal jaw apparatus. In R. Shadwick and G.V. Lauder (eds.), *Biomechanics of Fishes*. San Diego: Academic Press, pp.77–101.

(2007). Functional versus morphological diversity in macroevolution. *Annual Review of Ecology, Evolution & Systematics* 38, 381–401.

Wainwright, P.C., Alfaro, M.E., Bolnick, D.I., and Hulsey, C.D. (2005). Many-to-one mapping of form to function: a general principle in organismal design? *Integrative and Comparative Biology* 45, 256–62.

Wainwright, P.C., Bellwood, D.R., Westneat, M.W., Grubich, J.R., and Hoey, A.S. (2004). A functional morphospace for the skull of labrid fishes: patterns of diversity in a complex biomechanical system. *Biological Journal of the Linnean Society* 82, 1–25.

Westneat, M.W. and Alfaro, M.E. (2005). Phylogenetic relationships and evolutionary history of the reef fish family Labridae. *Molecular Phylogenetics and Evolution* 36, 370–90.

Wright, S. (1932). The roles of mutation, inbreeding, crossbreeding and selection in evolution. *Proceedings of the XI International Congress of Genetics* 1, 356–66.

7

The developmental evolution of avian digit homology: an update

GÜNTER P. WAGNER

INTRODUCTION

It is a truism of evolutionary developmental biology that the evolution of phenotypic characters has to be caused by the evolutionary modification of their developmental pathways (Hall 1998; Raff 1996). Surprisingly, however, the evolutionary conservation of a phenotypic character does not imply the conservation of its developmental pathway (Hall 1994; Wagner and Misof 1993; Weiss and Fullerton 2000). A paradigmatic case of this problem is the homology of the digits in the bird wing. Phylogenetic evidence conclusively shows that the digits of the bird wing are the thumb, the index finger, and the middle finger – that is, digits DI, DII, and DIII (Sereno 1999b). The embryological origin of these fingers, however, is identical to that of those fingers that usually form the index, middle, and "ring" finger – that is, digits DII, DIII, and DIV (Burke and Feduccia 1997; Hinchliffe and Hecht 1984; Müller and Alberch 1990). This discrepancy has been unresolved since the discovery of the dinosaur affinities of birds in the mid nineteenth century, and it continues to fuel a heated debate among paleontologists, ornithologists, and developmental biologists (Feduccia 1996, 1999, 2001; Galis *et al.* 2003; Prum 2002). Furthermore, it created a discrepancy in the naming of the wing digits between the evolutionary literature and the considerable number of

Research on avian digit identity is financially supported by NSF grant (IBN 0445971). I thank Ron Amundson, Alex Vargas, Frietson Galis, Mihaela Pavlicev, and Rasmus Winther, as well as the members of the my lab, for discussions on the subject of this paper. I also want to thank Professor Vincenzo Caputo for his help in collecting *Chalcides chalcides* specimens. This essay originally appeared as Wagner, G.P. (2005). The developmental evolution of avian digit homology: an update. *Theory in Biosciences* 124, 165–83, included here with kind permission of Springer Science and Business Media.

papers in molecular developmental biology, because the chick wing is an important model system. To settle this question would thus facilitate the integration of evolutionary and developmental biology into the nascent field of evolutionary developmental biology, because the acid test for interdisciplinary integration is the ability of researchers from different fields to agree on the relevant evidence and the inferences made from it (Nyhart 1995).

In the last ten years, the amount and the quality of evidence regarding the question of avian digit homology has markedly increased, due to novel paleontological finds and new molecular developmental data. For this reason, it may be useful to summarize these new data and to evaluate the conflict in the light of this. It will be argued that the preponderance of evidence points towards the possibility of a digit identity frame shift, which led to a dissociation between digit identity and embryological origin of the digits.

FOUR WAYS TO RESOLVE THE CONFLICT

Above, I have argued that the question of avian digit homology is essentially a conflict between two kinds of evidence associated with two ways of making scientific inferences. On the one hand, there is comparative biology, consisting of phylogenetics, comparative anatomy, and paleontology, which use the comparison of adult structures and methods of phylogenetic inference to identify the avian digits as digits DI, DII, and DIII. On the other hand, there is developmental biology, using comparisons of early developmental stages to infer digit homology. Until very recently, the developmental evidence overwhelmingly supported the hypothesis that the avian digits are digits DII, DIII, and DIV. This state of affairs is quite distressing, given that the inability to resolve similar conflicts spelled the end of an otherwise successful scientific paradigm at the end of the nineteenth century, namely that of evolutionary morphology (Nyhart 2002). Letting conflicts like these fester convinces students and academic administrators that the field is intellectually bankrupt, with dire consequences for its future. We are currently experiencing a renaissance of interest in the connection between development and evolution (Carroll *et al.* 2001; Hall 1998). The question thus is whether, at the beginning of the twenty-first century, we now have the technical and conceptual means to overcome the conflicts which haunted biologists more than a century ago, and whether we succeed in establishing conceptual continuity between evolutionary and developmental biology (Amundson 2005). I think that avian digit homology is a paradigmatic problem for evolutionary developmental biology to solve in its new molecular incarnation.

Given that the avian digit homology problem results from a conflict between two ways of making scientific inferences, it is natural to seek to resolve the issue in two principal ways: first, by questioning the inferences made from the anatomical and paleontological data, and second, by questioning the inferences from the developmental data. Within each category there are currently two ideas being discussed, which adds up to four possibilities that we need to consider.

Questioning the inference from paleontological data:

- Birds are not derived from theropod dinosaurs (Feduccia 1996).
- The digits in the hands of theropods are digits DII, DIII, and DIV, but have assumed the morphology of digits DI, DII, and DIII (Pyramid Reduction Hypothesis)(Galis *et al.* 2003; Kundrát *et al.* 2002).

The first hypothesis is part of a larger controversy about the evolutionary origin of birds. It is fair to say, however, that the theropod derivation of birds is no longer seriously in doubt. Overwhelming evidence emerging from fossil finds over the last ten years, in particular the discovery of feathered dinosaurs, has essentially settled the question (Prum 2002; Sereno 1999b). In this paper, I will not consider the idea that birds are anything else but theropod dinosaurs. More recently, a more subtle point has been raised by Frietson Galis and colleagues (2003) that requires closer attention. It has to do with the phylogenetic position of *Eoraptor* and *Herrerasaurus* which will be considered in the next section. This observation has been advanced to support the so-called Pyramid Reduction Hypothesis (Kundrát *et al.* 2002), which posits that the digits DI and DV were reduced, and that digits DII and DIV then assumed the phalangeal formulae of digits DI, DII, and DIII.

Questioning the inferences from developmental data:

- The timing of digit development has changed (Axis Shift Hypothesis =ASH) (Chatterjee 1998; Garner and Thomas 1998; Shubin 1994).
- Digit identity is dissociated from the location of the digit anlage (Frame Shift Hypothesis=FSH) (Wagner and Gauthier 1999).

In embryological studies, the most frequently used landmark to identify digit anlagen is the location of the first developing digit. This anlage usually develops into digit DIV (Shubin and Alberch 1986), and,

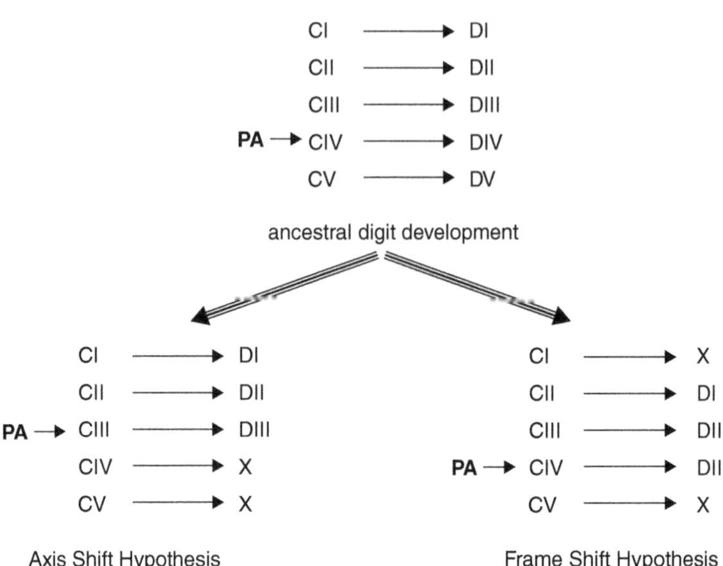

Figure 7.1 Schematic comparison of ancestral digit development (top) and the modifications to avian digit development, according to the Axis Shift Hypothesis (ASH) and the Frame Shift Hypothesis (FSH). In the ancestral digit development, the primary axis (PA) gives rise to the condensation CIV and the five condensations CI to CV develop into digits of corresponding identities DI to DV. According to the ASH, the digits DI, DII, and DIII still develop from condensations CI, CII, and CIII, but the primary axis has shifted to give rise to CIII. This hypothesis predicts that in birds, two posterior condensations do not develop into definite digits. According to the FSH, the primary axis remains where it was in the ancestral state, but the three digits in the bird wing develop from condensations CII, CIII, and CIV. This hypothesis predicts that an anterior and a posterior condensation fail to develop into definite digits. These predictions have been tested with embryological studies that visualize pre-chondrogenic condensations, showing that there is an anterior condensation that does not give rise to a definite digit (Feduccia and Nowicki 2002; Kundrát *et al.* 2002; Larsson and Wagner 2002; Welten *et al.* 2005).

consequently, the digits developing anterior to the earliest developing digit are identified as digits DI, DII, and DIII. To my knowledge, Shubin (1994) was the first to question this inference, suggesting that it is possible that there was heterochrony in the sequence of digit development, such that in theropods the anlage of digit DIII develops first with progressive reduction of the anlage of digits DIV and DV. Chatterjee (1998), as well as Garner and Thomas (1998), elaborated this scenario (fig. 7.1) in response to the paper by Burke and Feduccia (1997), using the primary axis as a landmark to identify the avian digits. This scenario has been called Axis Shift Hypothesis (ASH) (Larsson and Wagner 2002). New developmental data regarding the Axis Shift Hypothesis will be discussed later in this chapter.

The last hypothesis discussed here assumes that the pattern of digit development has not changed – that is, that the primary axis in the bird wing is homologous to that in other amniotes, but that the developmental fate of the digit condensations has changed, such that condensation CII develops into digit DI, and condensation CIII into digit DII, and so on. (Wagner and Gauthier 1999) (fig. 7.1). This scenario has been called Frame Shift Hypothesis (FSH) because it assumes a shift in the spatial relationship between adult digit identity and digit condensations.

EVIDENCE REGARDING THEROPOD DIGIT IDENTITY

There are two main steps in the argument that leads to the identification of the bird digits as DI, DII, and DIII. The first is the great similarity between the hand of *Archaeopteryx* and that of certain theropod dinosaurs, which are more closely related to *Archaeopteryx* and crown-group birds than to other theropods. The similarities include the phalangeal formula of 2, 3, and 4, which is the same as the plesiomorphic state for Tetrapoda in general, as well as the elongated penultimate phalange, the sharply pointed prehensile claws, and numerous characters concerning the shape of each element and their articulation surfaces (Gauthier 1986; Sereno 1993). These similarities imply that the digits of birds are those of the three functional digits of the non-avian maniraptorians (which is a clade in the theropods that includes the birds). The second step is the identification of the theropod digits as DI, DII, and DIII, based on the similarity of the first metacarpal of theropods and that of basal pentadactyl dinosaurs, which have an offset head that directs the first digit away from the other fingers (Gauthier 1986), and the fact that the most basal theropod lineages, *Eoraptor* and *Herrerasaurus*, have two metacarpals posterior to the three functional fingers. These have at most one or two phalanges that have not protruded from the palm. Furthermore, the trend to reduce digits IV and V is already visible in crocodiles, where digits I, II, and III are much stronger than digits IV and V. This pattern is also seen in some non-theropod dinosaurs, like *Heterodontosaurus tucki* (Ornithischia), the sauropodomorphs *Thecodontosaurus antiquus*, *Efraasia diagnostica*, and *Massospondylus carinatus* (Gauthier 1986). This supports the hypothesis that the digits DIV and DV have been reduced in theropods. Galis and colleagues (2003), however, argue that the phylogenetic relationships among basal dinosaur lineages are not clear, and that there remains the possibility that the basal theropods have lost digits DI and DV, rather than digits DIV and DV. If this were indeed the case, the basis for the theropod digit homology would be based only on the special structure of the most anterior finger (=thumb).

The phylogenetic position of the critical taxa, *Eoraptor* and *Herrerasaurus*, has been debated (Gauthier 1986; Holtz 1995; Padian 1992; Padian and May 1993; Sereno 1999a, b). While *Eoraptor* was described only in 1993 (Sereno *et al.* 1993), *Herrerasaurus* has been known from fragmentary evidence since 1963 (Reig 1963). Originally, *Herrerasaurus* was classified as a theropod (Benedetto 1973; Colbert 1970; Reig 1963), but this classification was later found to be based on plesiomorphic characters (Novas 1997). In a cladistic re-analysis of the data, this taxon was reconstructed as a sister group to the remaining dinosaurs (Gauthier 1986). After the discovery of a more complete specimen, the phylogenetic position of *Herrerasaurus* was changed back again to the sister taxon to all remaining theropods (Novas 1993; Sereno 1994; Sereno and Novas 1992). The reason for this conflict between earlier and later phylogenetic reconstructions is that the twelve synapomorphic characters that unite *Herrerasaurus* with the remaining theropods have not been known until the more complete material was discovered (Novas 1993; Sereno 1993; Sereno 1994; Sereno and Novas 1992; Sereno *et al.* 1993). Hence the discrepancies in the literature are largely due to the incompleteness of the original specimen.

Another difficulty was that *Herrerasaurus* has an apparently plesiomorphic character state in the attachment of the pelvic girdle to the spine (two vertebrae instead of five), while both the Ornithischia and the remaining theropods have more than two. In the light of the novel data, this character is either interpreted as a reversal (Novas 1993) or a convergence, because different vertebrae are integrated into the sacrum in the Ornithischia and the Saurischia (Gauthier personal communication).

For *Eoraptor*, the situation is more difficult, because it seems to belong to a lineage even more basal than that of *Herrerasaurus* and thus has fewer known synapomorphies with the theropods. Sereno and collaborators (Sereno *et al.* 1993) list five synapomorphies, of which three are hand characters. These hand characters could be considered problematic for the question considered here, given that we want to assess their homology. Among the remaining two synapomorphies, there is one – "extreme hollowing of centra and long bones" – that has been considered problematic (Novas 1993), due to wide distribution of this character state in dinosaurs. Hence, the *Eoraptor* affiliation is based mostly on hand characters and is problematic in the present context, where digit homology is in question.

The problem with the phylogenetic position of *Eoraptor*, however, is not critical to our argument, because the affiliation of *Herrerasaurus* to theropods is supported by twelve synapomorphies excluding many problematic characters, of which only four are hand characters (Novas 1993). Hence, there are at least eight synapomorphies unrelated to hand morphology supporting the theropod nature of *Herrerasaurus*. Since *Herrerasaurus* clearly has two posterior metacarpals, the identity of the functional three digits of theropods is clearly DI, DII, and DIII. In conclusion, the homology of the theropod fingers is not seriously in doubt.

EVIDENCE REGARDING THE AXIS SHIFT HYPOTHESIS

As explained above, the ASH assumes a heterochronic change in the sequence of digit development in that, in the bird hand, the first digit to develop is not digit DIV but digit DIII, and therefore the "primary axis" no longer identifies the anlage of digit DIV but that of DIII (Chatterjee 1998; Garner and Thomas 1998; Shubin 1994). The main difficulty in evaluating this hypothesis is that for most of the twentieth century, embryologists only identified four digit anlagen in the bird wing (Burke and Feduccia 1997; Hinchliffe 1977; Hinchliffe and Hecht 1984), of which the three anterior ones form the definite digits of the bird hand. The location of the primary axis in the bird hand relative to the canonical five digits was not clear.

Embryological studies of limb skeletons in the second half of the twentieth century were dominated by two staining techniques – the incorporation of radioactive sulfate (Hinchliffe and Griffiths 1983) and staining with Alcian blue (Burke and Feduccia 1997; Müller and Alberch 1990). Both techniques detect the extracellular matrix produced during chondrification. Hence, these techniques identify regions of the embryo in which cartilages differentiate. It is, however, clear that the very first stage of skeletal development in the tetrapod limb is not cartilage differentiation (in contrast to the situation in paired ray-finned fish; Grandel and Schulte-Merker 1998), but the condensation of pre-chondrogenic cells (Hall and Miyake 1995; Shubin and Alberch 1986).

In a leap of interpretation, however, skeletal elements identified with sulfate incorporation and Alcian blue staining used to be identified with pre-chondrogenic condensations, which, by the nature of the staining technique, they could not be, namely pre-chondrogenic. Three different techniques have been used in recent years to overcome this technical limitation.

The first was the work by Kundrát and collaborators from Prague University (Kundrát and Seichert 2001, 2002), who used the fact that capillary blood vessels degenerate as pre-chondrogenic cells condense to form the first physical precursor of a skeletal element (Seichert and Rychter 1972). This phenomenon is caused by the production of an intrinsic angiogenesis inhibitor by the condensing cells (Hiraki and Shukunami 2000). Using India ink injections, Kundrát and colleagues were able to demonstrate a fifth zone of capillary regression, in addition to the four expected ones. The fifth capillary regression zone, indicating a possible pre-chondrogenic condensation, is located anterior to the four known digit anlagen (Kundrát *et al.* 2002).

Larsson and Wagner at Yale (2002) used another technique to visualize pre-chondrogenic condensations, the affinity of condensing cells to Peanut Agglutinin, a lectin specifically binding Galß1,3-Gal3NAc residues on glycoproteins (Dunlop and Hall 1995; Zschabitz 1998). With this technique, we also found evidence for an additional pre-chondrogenic condensation anterior to the three condrifying digit anlagen (fig. 7.2).

Most recently, Welten and collaborators at Leiden University (2005) used *in situ* hybridization to investigate the expression of *Sox9*, a transcription factor gene expressed in condensing skeletogenic cells, to map pre-chondrogenic condensations. They also found evidence for an additional anterior condensation, confirming the findings of Kundrát *et al.* (2002) and Larsson and Wagner (2002).

Finally, a different approach was used by Feduccia and Nowicki at the University of North Carolina (Feduccia and Nowicki 2002), who reasoned that larger species may proceed further in the development of their rudiments and chondrify what in the chick remains completely precartilaginous. They investigated the development of the wing of the ostrich with standard Alcian blue staining and found a chondrified protuberance anterior to the three definite digits, consistent with the findings cited above (Feduccia and Nowicki 2002). In summary, one has to conclude that the wings of birds (chickens and ostriches) contain five digit condensations of which in the chicken embryo four chondrify and three ossify.

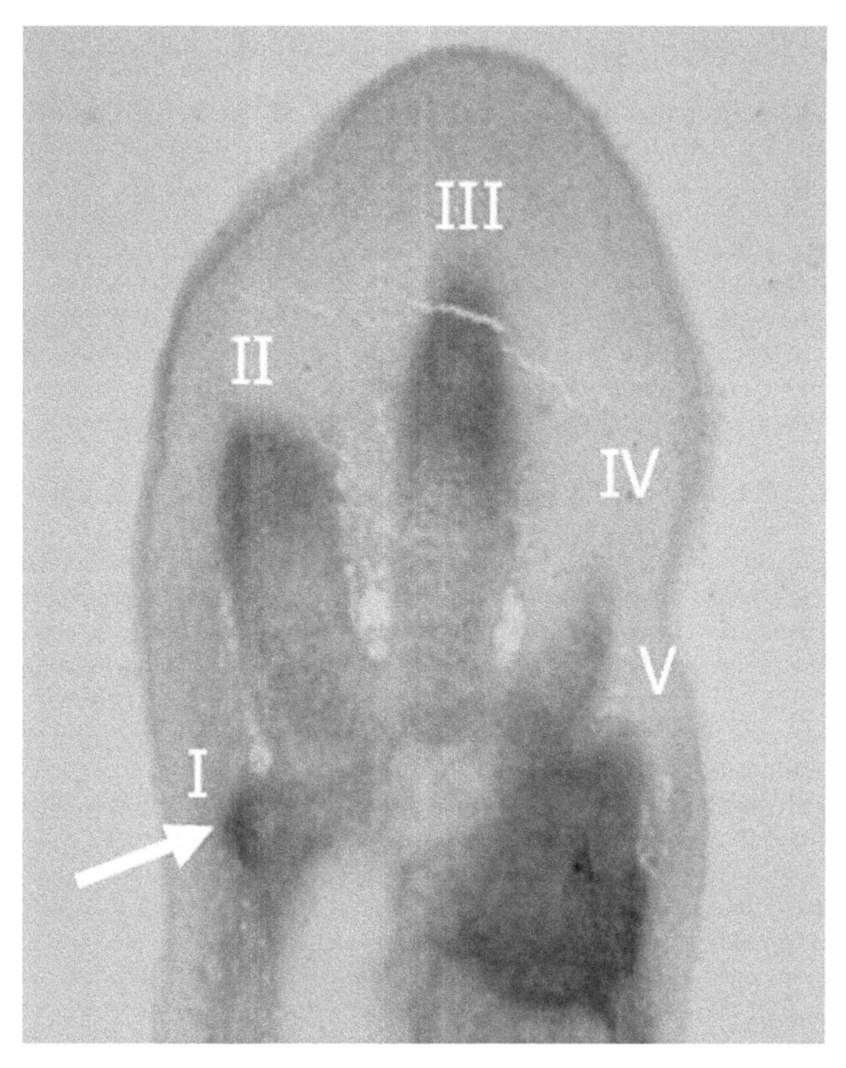

Figure 7.2 Section through the chicken wing bud, visualizing condensing cells with HRP-conjugated Peanut Agglutinin (H. Larsson unpublished data, reproduced with permission). Note that there are five centers of condensation with the three middle ones developing into digits. Hence, there are two condensations, one anterior and one posterior, that do not develop into definite digits. This result is inconsistent with the Axis Shift Hypothesis, but consistent with the Frame Shift Hypothesis.

The significance of this finding is that it gives a definite answer to the question whether the primary axis in the bird wing is different from that in other amniotes. If there were an anterior shift of the primary axis (CIII appears first instead of CIV), as hypothesized by Shubin, Chatterjee, Garner, and Thomas, one would predict that there are two condensations posterior to the primary axis and two anterior. If, however, the primary axis is not shifted there should be three condensations anterior to the primary axis and one posterior, as in all other amniotes investigated (Burke and Alberch 1985; Müller and Alberch 1990; Shubin and Alberch 1986) (fig. 7.1). The five papers cited above clearly show that there are three condensations anterior to the primary axis, implying that it has not changed in theropod phylogeny. This effectively falsifies the ASH.

While the overall picture in this area is quite clear, there are a few minor issues which should be investigated to test the robustness of the conclusion about ASH. An alternative interpretation of the additional pre-chondrogenic condensation found in recent years is the possibility that it is not the ancestral anlage of digit DI but a pre-pollex – that is, an additional digit anlage – that sometimes leads to a spontaneous additional digit anterior to digit DI (Braus 1906), and is found in many anuran species (Fabrezi 2001). A critical test of this possibility would be to investigate whether a pre-chondrogenic pre-pollex exists in the limb bud of crocodilians, the closest living relatives of birds. This should be possible with the same techniques that have been used in the chick wing. Another issue is the nature of an additional condensation at the posterior limit of the limb bud, which is usually interpreted as the anlage of a pisiform bone. Some authors have speculated that this could be a digit rudiment (Vargas and Fallon 2005; Welten *et al.* 2005). It would be worthwhile to clarify the nature of this condensation.

EVIDENCE REGARDING THE FRAME SHIFT HYPOTHESIS

The FSH differs from other interpretations of the evidence by rejecting the notion that digit identity is necessarily linked to the location of the digit anlage. The inspiration for this idea derives from the extensively documented fact that organ identity in many characters is not rigidly determined by the embryological origin of the character (Hall 1994). Radical forms of character identity transformations have been known for more than a century and are called homeotic transformations. In evaluating the FSH, however, we need to ask whether the dissociation between digit origin and digit identity has been specifically documented.

There is much evidence from the genetic and the experimental literature (Hinchliffe and Johnson 1980) that both digit number and digit morphology can be dissociated. I do not want to attempt a comprehensive list of results to support this conclusion, but only to point to a particularly dramatic mutation, the phenotype of knockout mutations of *Gli3*. These mutants have super-numerous digits of identical morphology, resembling digit DI (Litingtung *et al.* 2002; Welscher *et al.* 2002). The anterior-posterior differentiation of the autopod depends on a signaling center, called the Zone of Polarizing Activity (ZPA), which produces *Shh* as a signaling molecule (Riddle *et al.* 1993). *Gli3* codes for a protein necessary for the transduction of *Shh* signaling. In a *Gli3* mutant, the *Shh* signal is not "read," and the morphology of all digits is identical. Clearly, the position of a digit per se does not determine digit morphology, but can be influenced by the gene regulation independent of position.

In this context, a rare human disorder, tri-phalangeal thumb (TPT), is of particular interest. A TPT is a malformation of the first digit of the human hand which can be associated with various other symptoms of variable pathogenic origin (Qazi and Kassner 1988) (fig 7.3A). Among the various kinds of TPT is a distinct form which is, in its strongest expression, a perfect homeotic transformation of the thumb into an index finger, called non-opposable tri-phalangeal thumb (no-TPT) (MIM#190600, from the Online Mendelian Inheritance in Man™: http://www.ncbi.nlm.nih.gov/entrez/query.fcgi?db=OMIM&itool=toolbar). The morphology of the no-TPT resembles that of the posterior digits 2 to 5, not only in the number of phalanges, but also in many other fundamental morphological characters (fig. 7.3). The metacarpal I (MC-I) is subequal to the MC-II and gracile, unlike the normal MC-I, which is short and robust (fig. 7.3B). In normal hands, the MC-I has a proximal epiphysis, while MCs II to V have a distal epiphysis (Hess 1957). The MC-I of the no-TPT has a distal epiphysis. The proximal joint of the MC-I in primates is saddle-shaped, allowing the characteristic opposability. The proximal joint of the no-TPT is like that of MCs II to V, and therefore not opposable. The pisiform bones normally present at the distal joint of the MC-I (fig. 7.3B) are absent in the no-TPT, and the characteristic muscles associated with the thumb are also absent (*Mm. abductor/ opponens/ adductor pollicis*). There is no question that no-TPT represents a homeotic transformation of digit identity, as first hypothesized

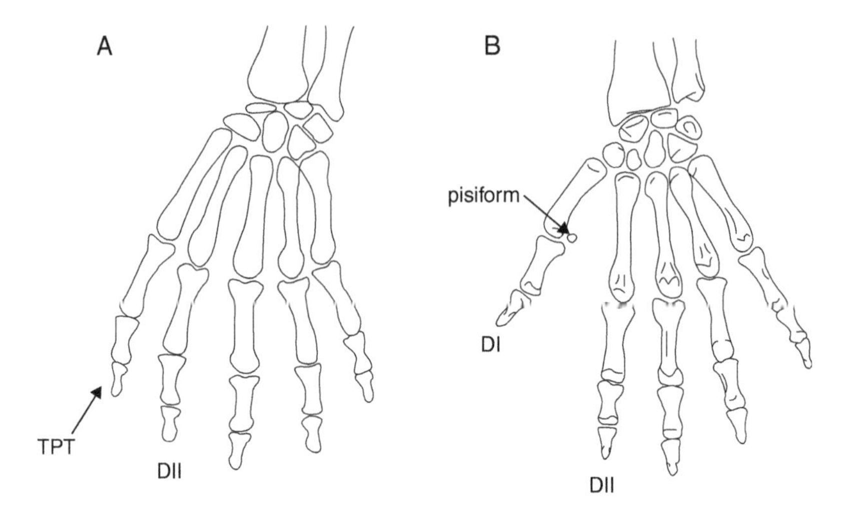

Figure 7.3 Drawings of skeletal elements from radiographs of the left human hand viewed from dorsal. A. The radiograph of a patient with a condition known as tri-phalangeal thumb (TPT) (from Hess, 1957). B. a radiograph of a normal hand. Note that the first digit in the TPT patient has many characteristics similar to the digit DII. It is likely that this condition is due to a homeotic transformation of DI into DII (Joachimsthal 1900; Warm *et al.* 1988). Drawing courtesy of Casey Dunn.

by Joachimsthal (1900) and confirmed by Warm and collaborators (1988). In 90 percent of cases, the no-TPT is found on both hands and inherited in a dominant autosomal mode (Hess 1957; Swanson and Brown 1962; Warm *et al.* 1988). Hence, there are genetic mutations which can change digit identity without compromising the mental ability or fertility of the affected individuals – otherwise, extensive families with this disorder, as studied by Swanson and Brown (1962) and Warm *et al.* (1988), would not exist. The patient described by Hess (1957) reported to have an advantage playing the piano, also indicating a normal mental capacity. It thus seems plausible that mutations affecting digit identity do not need to have widespread pleiotropic effects and so might be fixed by natural selection.

Another piece of evidence relevant to the FSH is that digit identity remains labile until late in development. Dahn and Fallon have shown that, in the chicken, digit morphology is influenced by signaling from the interdigital mesenchyme (IDM) posterior to the developing digit (Dahn and Fallon 2000). Manipulations of the IDM affect digit morphology long after the metacarpals have chondrified – that is, long after the stage of development that is used as embryological evidence for digit identity by embryologists. This implies that the patterning of digit number and digit identity are temporally dissociated and may thus also evolve independently.

Vargas and Fallon (2005) have used *Hox* gene expression patterns to test the FSH. They observed that in late stages of development, the digit DI is distinguished from the other digits by expression of *HoxD13*, but the absence of *HoxD12* gene expression, while the digits DII to DV express both genes. This was found to be the case in the chick hindlimb and in the mouse, where digit identity is not in question. They further tested the association between *Hox* gene expression and digit identity in the Silkie mutant in chicken, where a digit DII develops anterior to the digit DI in the foot. Again, DI was distinguished by the presence of *HoxD13* and the absence of *HoxD12* expression from the rest of the digits, even though, in this mutant, digit DI is not the most anterior digit in the limb. Hence, this expression pattern is not just a marker of the anterior edge of the limb bud, but could be specific to digit DI, at least in mouse and chick hindlimb. They then reported the expression pattern in the wing, finding that the anterior most definite digit expresses *HoxD13* but not *HoxD12*, consistent with the hypothesis that this digit is in fact DI (see fig. 7.4). This is the first developmental evidence consistent with the paleontological data on digit identity, and also suggests that changes in the *Hox* gene expression patterns specific to the hand could have effected the digit identity shift, as proposed by the FSH (Wagner and Gauthier 1999). The phylogenetic inference has recently been confirmed in a study of *HoxD11* expression (which in the mouse has the same expression pattern as *HoxD12* in *Alligator*). It was shown that the anterior limit of *HoxD11* expression in the alligator hand is one digit position more anterior than in the chicken wing (Vargas *et al.* 2008). This confirms that the posterior position of *HoxD11/12* expression in the bird wing is phylogenetically derived, as predicted by the FSH.

CONFLICT BETWEEN ANATOMICAL AND EMBRYOLOGICAL EVIDENCE ON DIGIT IDENTITY IN OTHER ANIMALS: *LO STRANO CASO DELLA LUSCENGOLA*

Digit reduction is a common mode of evolution and has been studied extensively since the advent of evolutionary theory (Fürbringer 1870).

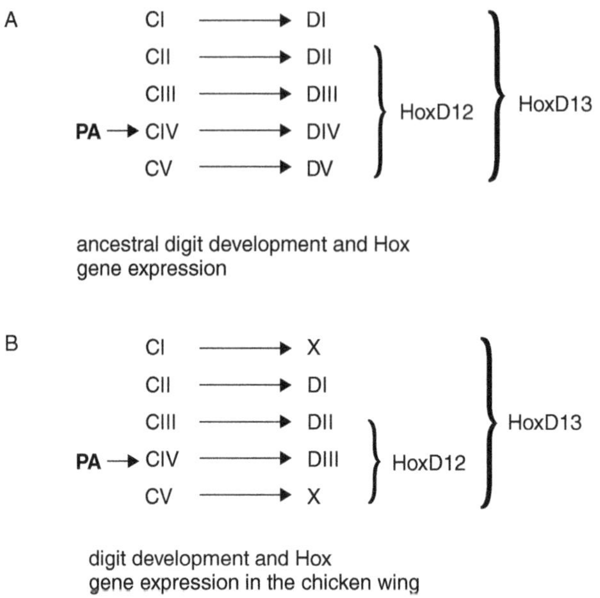

A

CI → DI
CII → DII
CIII → DIII
PA → CIV → DIV
CV → DV

HoxD12 (DII–DV)
HoxD13 (DI–DV)

ancestral digit development and Hox gene expression

B

CI → X
CII → DI
CIII → DII
PA → CIV → DIII
CV → X

HoxD12 (DII–X)
HoxD13 (X–X)

digit development and Hox gene expression in the chicken wing

Figure 7.4 Schematic comparison of *Hox* gene expression in the ancestral hand (A) and the bird wing (B), according to Vargas and Fallon (2005). In mice hands and bird hindlimb and the alligator hand (Vargas *et al.* 2008) representing the likely ancestral condition, the digit which develops from condensation CI expresses *HoxD13* but not *HoxD12*, while the digit DII expresses both *HoxD12* and *HoxD13*. In the chick wing, however, the digit developing from the condensation CII expresses *HoxD13* but not *HoxD12*. This is consistent with the Frame Shift Hypothesis, which proposes that, in the bird wing, the digit developing from condensation CII assumes the adult identity of digit DI, and the digit developing from CIII assumes the identity of digit DII.

An interesting taxon for the reconstruction of the stepwise reduction of digits and limbs to the complete loss of these structures are the skinks (Greer 1991). Within the skinks, at least one case is known where a similar conflict exists between anatomical and embryological evidence as in birds, namely the Italian three-toed skink, or *Luscengola* (*Chalcides chalcides*) (Orsini and Cheylan 1981). This case has not been discussed in the most recent literature, maybe because the primary literature on this problem is entirely in German and French (see below). I want to summarize the relevant results here for the first time in English, to make this fascinating segment of the scientific literature more accessible.

Ch. chalcides has three well-formed digits in both the forelimb and the hindlimb, with a rudimentary metatarsal posterior to the three. Based on anatomical evidence, Fürbringer (1870) (in his time, *Chalcides*

chalcides was called *Seps tridactylus*) identified the three fully formed digits as digits DI, DII, and DIII, a conclusion confirmed by other investigators looking at the structural similarities between *Chalcides chalcides* limbs and those of species with five digits, like *Chalcides ocellatus* (Renous-Lecuru 1973; Steiner and Anders 1946). In contrast, all the embryological studies agree that the three digits of *Luscengola* develop from anlagen that normally give rise to digits DII, DIII, and DIV in the lizards with five digits (Raynaud *et al.* 1986; Sewertzoff 1931) – that is, these digits develop from condensations CII, CIII, and CIV. This is exactly the same situation as in avian digits.

Within the taxon *Chalcides*, most species have either five or three digits (Caputo *et al.* 1995; Greer *et al.* 1998), with the exception, of two species with four digits, *Ch. mionecton* and *Sphenops sphenopsiformis*, which is nested within the taxon *Chalcides* according to a new phylogenetic analysis of skinks (Brandley *et al.* 2005). These two species have a rudimentary metapodial (MP, being either a metacarpal or a metatarsal) MP-V, showing that the reduction in these species is clearly postaxial, unlike the tendency in mammals, where digit DI tends to be reduced first. Unfortunately, there is no species of *Chalcides* described that, like *Herrerasaurus* in theropods, has three fully formed digits and two metapodials in the adult skeleton. The position of these rudimentary metapodials would be very interesting in terms of the phylogenetic history of digit loss in this taxon.

The discrepancy between the expected pattern of digit reduction based on mammalian examples and the results from *Luscengola* did not escape the attention of the investigators, starting with Fürbringer (1870). In this context, an interesting point was raised by Steiner and Anders (1946). These authors noted that one needs to distinguish between various forms of digit loss. On the one hand, there is the case of rudimentation, basically a melting away of the limb when it loses its function completely. But Steiner and Anders point out that the limb of *Luscengola* is not functionless; in particular, the three digits are not reduced, but well formed. The limbs are used to balance the body when the animal is at rest (fig. 7.5), and to push through vegetation when it is moving slowly (Bruno and Maugeri 1976; Orsini and Cheylan 1981). The limbs, though small, are also used for burrowing in those *Chalcides* species that live in sandy soils. That also explains why the three digits of the *Luscengola* are fully formed and strong, not reduced except in size. Steiner and Anders (1946) propose that the digit loss in *Ch. chalcides* is not a case of rudimentation, but is due to an adaptive modification of

Figure 7.5 *Chalcides chalcides* at rest. Note that the small limbs are used to stabilize the body at rest. Hence, the much-reduced limbs are not functionless, and have been reported to be used also during slow crawling through grass.

the limb during which some digits get lost. Steiner and Anders call that mode of evolution "adaptive reduction" (p. 545). Note that in theropods also, digit loss is not a case of rudimentation *sensu* Steiner and Anders, because it happens during an adaptive modification of the hand for prey capture (Sereno 1999b; Wagner and Gauthier 1999). Hence, it is possible that in the taxon *Chalcides* we witness a neontological case, or digit identity shift, similar to the one in avian digits. The difference, though, is that in the case of *Chalcides* we have many closely related species with five, four, and three digits in the recent fauna.

FACTS WITHOUT THE FORCE OF EVIDENCE

In the previous sections I discussed recent empirical findings pertaining to the question of avian digit homology. The issues raised in these sections do not cover all the arguments that have been fielded in this debate. I have not discussed these remaining arguments because they differ from the results summarized above, in that they do not have the force of evidence in this context. They are questions about patterns of digit reduction and mechanistic plausibility (see below), and highlight important issues to be addressed in future research, but do not have the scientific standing to decide the problem at hand, namely theropod/avian digit identity. I am quick to acknowledge that it is not easy to say exactly what evidence is. But it is also clear that making the distinction between evidence and non-evidence is important for any science. For that reason, I want to start this section with a short account of how I make the distinction between evidence and non-evidence.

Evidence consists of empirical facts, together with a well-established theoretical framework, that allow scientific inferences. Furthermore, the empirical facts need to pertain directly to the question at hand, in this case avian digit homology – that is, observations of birds and their close relatives, the non-avian archosaurs. An obvious example is fossil data, which are used with accepted phylogenetic inference methods to establish the phylogenetic relationships of *Eoraptor*, *Herrerasaurus*, and the other dinosaurs. These data count as evidence because the osteology of the hand of these animals is directly relevant for determining the homology of the fully formed digits of theropods, and, by implication, of birds.

In contrast, facts that do not have the force of evidence are at least of two kinds. On the one hand, there are empirical generalizations that do not directly address the question of avian digit homology. The most often raised fact of this kind has to do with patterns of digit reduction in other taxa, in particular lizards and mammals (Kundrát *et al.* 2002; Montagna 1945; Müller and Alberch 1990; Raynaud and Clergue-Gazeau 1986; Sewertzoff 1931; Steiner 1934). However, findings on mammals, for example, cannot falsify hypotheses regarding theropods. There is no reason to believe that the generalizations from digit reduction patterns in one taxon – that is, mammals – have to apply to another taxon. The perceived difference might as well be real, and it is important to understand the biological reasons for these differences, if they exist; but one cannot argue that something did not happen in taxon A because it did not happen in taxon B. This certainly cannot be considered as evidence for the problem at hand.

The second type of argument, which cannot be admitted as evidence, is claims of impossibility based on incomplete or fragmentary knowledge. In the context of avian digit identity, these arguments come in two flavors. One is that a digit identity frame shift is implausible (read "impossible"), given our current mechanistic knowledge (Feduccia 1999; Galis *et al.* 2003). The other pertains to mechanisms of evolutionary change, claiming that the inferred changes in digit development cannot be true because we do not know what the adaptive advantage of them might have been (Galis *et al.* 2003, 2005). Both arguments point to true facts, namely that our mechanistic knowledge is incomplete and there is no mechanistic explanation of what we see in bird evolution – and it is true that there is no adaptive scenario which explains the differences in digit development between birds and alligators. But these facts are clearly irrelevant, because one is derivative of a very weak knowledge base and the other is only negative. Even though our knowledge of limb development has improved dramatically in the last twenty years, we still do not know the developmental-genetic limits of limb evolution. An inference cannot be stronger than the mechanistic knowledge from which it extrapolates. A historical example that shows the dangers of this type of argument is the history of continental drift and plate tectonics. The idea that continents are not nailed to the surface of the earth, proposed by Alfred Wegener in 1912, was largely based on fossils found in various parts of the world, the shape of continents, and similarities between rock formations on different continents. The greatest weakness of his theory was its mechanistic explanation. Wegener assumed that continents are pieces of the earth crust "swimming" on a sea of magma, and driven by centrifugal, Coriolis, and convection forces. Geophysicists were quick to point out that this is physically impossible, and they were right. What was fallacious, however, was the conclusion that for that very reason the continents did not move at all, and thus the continental drift hypothesis is wrong (Schwarzbach 1980). Clearly, what was wrong was Wegener's *explanation* of the phenomenon, and not the hypothesis that continental drift has occurred and is still occurring. Similarly, the question of how digits and digit development evolved in theropods is one that needs to be considered independently of whether we have the knowledge and the tools to find out what the mechanisms were that caused that change. For instance, it is not controversial that organ rudimentation happens in evolution, but we still do not know whether it is caused by mutation accumulation and genetic drift or by natural selection. A failure to explain what happened has no implications for whether an event occurred or not. This failure only reflects the limits of our current knowledge.

Of similar structure is the argument that FSH cannot be correct because there is no adaptive reason for this change to occur (Galis *et al.* 2003). But again, this argument is one of a lack of explanation, not one of relevant empirical evidence. A historical fact can be true whether we can explain it or not. Hence, one should first focus on the question of which hypothesis of fact – ASH, FSH, and so on – is correct, and only then focus on how to explain it. The sum of current evidence is pointing toward the possibility of a digit identity frame shift in the stem lineage of birds, and possibly also within *Chalcides* skinks. Furthermore, it is possible that this change was caused by a change in the expression of *Hox* genes (Vargas and Fallon 2005; Vargas *et al.* 2008). Now the challenge is to find out whether this mechanistic *explanation* is true, and what evolutionary forces may have caused the digit identity frame shift.

REFERENCES

Amundson, R. (2005). *The Changing Role of the Embryo in Evolutionary Thought: Roots of Evo-Devo*. Cambridge University Press.

Benedetto, J.L. (1973). Herrerasauridae, nueva familia de sarisquios triscos. *Ameghiniana* 10, 89–102.

Brandley, M.C., Schmitz, A., and Reader, T. (2005). Partitioned Bayesian analyses, partition choice, and the phylogenetic relationships of scincid lizards. *Syst Biol* 54, 373–90.

Braus, H. (1906). Die Entwicklung der Form der Extremitäten und des Extremitätenskeletts. In O. Hertwig (ed.), *Handbuch der vergleichenden und experimentellen Entwicklungslehre der Wirbeltiere*. Jena: Gustav Fisher, vol. III, part 2, pp.167–338.

Bruno, S. and Maugeri, S. (1976). *Rettili d'Italia: Tartarughe e Sauri*. Florence: I.A. Martello.

Burke, A.C. and Alberch, P. (1985). The development and homology of the chelonian carpus and tarsus. *J Morphol* 186, 119–31.

Burke, A.C. and Feduccia, A. (1997). Developmental patterns and the identification of homologies in the avian hand. *Science* 278, 666–8.

Caputo, V., Lanza, B., and Palmieri, R. (1995). Body elongation and limb reduction in the genus *Chalcides* Laurenti 1768 (Squamata Scincidae): a comparative study. *Trop Zoology* 8, 95–152.

Carroll, S.B., Grenier, J.K., and Weatherbee, S.D. (2001). *From DNA to Diversity*. Malden, MA: Blackwell Science.

Chatterjee, S. (1998). Counting the fingers of birds and dinosaurs. *Science* 280, 355a.

Colbert, E. (1970). A saurischian dinosaur from the Triassic of Brazil. *American Museum Novitates* 2405, 1–39.

Dahn, R.D. and Fallon, J.F. (2000). Interdigital regulation of digit identity and homeotic transformation by modulated BMP signaling. *Science* 289, 438–41.

Dunlop, L.-L.T. and Hall, B.K. (1995). Relationships between cellular condensation, preosteoblast formation and epithelial-mesenchymal interactions in initiation of osteogenesis. *Int J Dev Biol* 39, 357–71.

Fabrezi, M. (2001). A survey of prepollex and prehallux variation in anuran limbs. *Zool J Linn Soc* 131, 227–48.

Feduccia, A. (1996). *The Origin and Evolution of Birds*. New Haven: Yale University Press.

(1999). 1,2,3=2,3,4: accommodating the cladogram. *Proc Natl Acad Sci USA* 96, 4740–2.

(2001). Digit homology of birds and dinosaurs: accommodating the cladogram. *Trends Ecol Evol* 16, 285–6.

Feduccia, A. and Nowicki, J. (2002). The hand of birds revealed by early ostrich embryos. *Naturwissenschaften* 89, 391–3.

Fürbringer, M. (1870). *Die Knochen und Muskeln der Extremitäten bei den schlangenähnlichen Sauriern: vergleichend anatomische Abhandlung*. Leipzig: Verlag von Wilhelm Engelmann.

Galis, F., Kundrát, M., and Metz, J.A.J. (2005). *Hox* genes, digit identities and the theropod/bird transition. *J Exp Zoolog B Mol Dev Evol* 304B, 198–205.

Galis, F., Kundrát, M., and Sinervo, B. (2003). An old controversy solved: bird embryos have five fingers. *Trends Ecol Evol* 18, 7–9.

Garner, J.P. and Thomas, A.L.R. (1998). Counting the fingers of birds. *Science* 280, 355.

Gauthier, J. (1986). Saurischian monophyly and the origin of birds. *Mem Calif Acad Sci* 8, 1–55.

Grandel, H. and Schulte-Merker, S. (1998). The development of the paired fins in the Zebrafish (*Danio rerio*). *Mechanisms of Development* 79, 99–120.

Greer, A.E. (1991). Limb reduction in Squamates: identification of the lineages and discussion of the trends. *J Herpetology* 25, 166–73.

Greer, A.E., Caputo, V., Lanza, B., and Palmieri, R. (1998). Observations on limb reduction in the scincid lizard genus *Chalcides*. *J Herpetology* 32, 244–52.

Hall, B.K. (1994). Homology and embryonic development. In M.K. Hecht, R.J. MacIntyre, and M.T. Clegg (eds.), *Evolutionary Biology*. New York: Plenum Press, vol. XXVIII, pp. 1–30.

(1998). *Evolutionary Developmental Biology*. London: Chapman and Hall.

Hall, B.K. and Miyake, T. (1995). Divide, accumulate, differentiate: cell condensation in skeletal development revisited. *Int J Dev Biol* 39, 881–93.

Hess, H. (1957). Beiderseitige kongenitale daumenlose Fünffingerhand bei Mutter und Kind. *Z Anat Entwicklungsgesch* 120, 226–31.

Hinchliffe, J.R. (1977). The chondrogenic pattern in chick limb morphogenesis: a problem of development and evolution. In D.A. Ede, J.R. Hinchliffe, and M. Balls (eds.), *Vertebrate Limb and Somite Morphogenesis*. Cambridge University Press, pp. 293–309.

Hinchliffe, J.R. and Griffiths, P.J. (1983). The prechondrogenic patterns in tetrapod limb development and their phylogenetic significance. In B.C. Goodwin, N. Holder, and C.C. Wylie (eds.), *Development and Evolution*. Cambridge University Press, pp. 99–121.

Hinchliffe, J.R. and Hecht, M. (1984). Homology of the bird wing skeleton. *Evol Biol* 20, 21–37.

Hinchliffe, J.R. and Johnson, D.R. (1980). *The Development of the Vertebrate Limb*. New York: Oxford University Press.

Hiraki, Y. and Shukunami, C. (2000). Chondromodulin-I as a novel cartilage-specific growth-modulating factor. *Pediatr Nephrol* 14, 602–5.

Holtz, T.R., Jr. (1995). A new phylogeny of the Theropoda. *J Vert Paleont* 15, 35A.

Joachimsthal, G. (1900). Verdoppelung des linken Zeigefingers und Dreigliederung des rechten Daumens. *Berliner Klinische Wochenschrift* 37, 835–8.

Kundrát, M. and Seichert, V. (2001). Developmental remnants of the first avian metacarpus. *J Morphol* 248, 252A.

Kundrát, M., Seichert, V., Russell, A.P., and Smetana, K., Jr. (2002). Pentadactyl pattern of the avian wing autopodium and pyramid reduction hypothesis. *J Exp Zool Mol Dev Evol* 294, 152–9.

Larsson, H.C.E. and Wagner, G.P. (2002). The pentadactyl ground state of the avian wing. *J Exp Zool Mol Dev Evol* 294, 146–51.

Litingtung, Y., Dahn, R.D., Li, Y., Fallon, J.F., and Chiang, C. (2002). *Shh* and *Gli3* are dispensable for limb skeleton formation but regulate digit number and identity. *Nature* 418, 979–83.

Montagna, W. (1945). A re-investigation of the development of the wing of the bird. *J Morphol* 76, 87–118.

Müller, G.B. and Alberch, P. (1990). Ontogeny of the limb skeleton in *Alligator mississippiensis*: developmental invariance and change in the evolution of Archosaur limbs. *J Morphol* 203, 151–64.

Novas, F.E. (1993). New information on the systematics and postcranial skeleton of *Herrerasaurus ischigualastensis* (Theropoda: Herrerasauridae) from the Ischigualasto formation (Upper Triassic) of Argentina. *J Vert Paleont* 13, 400–23.

(1997). Herrerasauridae. In P.J. Currie and K. Padian (eds.), *Encyclopedia of Dinosaurs*. San Diego: Academic Press, pp. 303–11.

Nyhart, L.K. (1995). *Biology Takes Form. Animal Morphology and the German Universities, 1800–1900*. University of Chicago Press.

(2002). Learning from history: morphology's challenges in Germany ca 1900. *J Morphol* 252, 2–14.

Orsini, J.-P. and Cheylan, M. (1981). *Chalcides chalcides* (Linnaeus 1758) – Erzschleiche. In W. Böhme (ed.), *Handbuch der Reptilien und Amphibien Europas*. Wiesbaden: Akademische Verlagsgesellschaft, vol. I, pp. 318–37.

Padian, K. (1992). A proposal to standardize tetrapod phalangeal formula designations. *J Vert Paleont* 12, 260–2.

Padian, K. and May, C.L. (1993). The earliest dinosaurs. *New Mexico Museum of Natural History and Science Bulletin* 3, 379–81.

Prum, R.O. (2002). Why ornithologists should care about the theropod origin of birds. *The Auk* 119, 1–17.

Qazi, Q. and Kassner, E.G. (1988). Triphalangeal thumb. *J Med Genet* 25, 505–20.

Raff, R. (1996). *The Shape of Life*. University of Chicago Press.

Raynaud, A. and Clergue-Gazeau, M. (1986). Identification des doigts réduits ou manquants dans les pattes des embryons de Lézard vert (*Lacerta viridis*) traités par la cystosine-arabinofuranoside. Comparaison avec les réductions digitales naturelles des espèces de reptiles serpentiformes. *Arch Biol (Bruxelles)* 97, 279–99.

Raynaud, A., Clergue-Gazeau, M., and Brabet, J. (1986). Remarques preliminaires sur la structure de la patte du Seps Tridactyle (*Chalcides chalcides*, L.). *Bull Soc Hist Nat, Toulouse* 122, 109–11.

Reig, O.A. (1963). La presencia de dinosaurios saurisquios en los "Estratos de Ischigualasto" (Mesotriásico superior) de las provincias de San Juan y La Rioja (República Argentina). *Ameghiniana* 3, 3–20.

Renous-Lecuru, S. (1973). Morphologie comparée du carpe chez les Lepidosauriens actuels (Rhynchocéphales, Lacertilens, Amphisbéniens). *Gegenbaurs morph Jahrb, Leipzig* 119, 727–66.

Riddle, R.D., Johnson, R.L., Laufer, E., and Tabin, C. (1993). Sonic hedgehog mediates the polarizing activity of the ZPA. *Cell* 75, 1401–16.

Schwarzbach, M. (1980). *Alfred Wegener und die Drift der Kontinente*. Stuttgart: Wissenschaftliche Verlagsgesellschaft.

Seichert, V. and Rychter, Z. (1972). Vascularization of developing anterior limb of the chick embryo II. Differentiation of vascular bed and its significance for the location of morphogenetic processes inside the limb bud. *Folia Morphologica (Warsz)* 19, 352–61.

Sereno, P. (1994). The pectoral girdle and forelimb of the basal theropod *Herrerasaurus ischigualestensis*. *J Vert Paleont* 13(4), 425–50.

Sereno, P.C. (1993). Dinosaurian precursors from the Middle Triassic of Argentina: *Lagerpeton chanarensis*. *J Vert Paleont* 13, 385–99.

(1999a). A rationale for dinosaurian taxonomy. *J Vert Paleont* 19, 788–90.

(1999b). The evolution of dinosaurs. *Science* 284, 2137–47.

Sereno, P.C., Forster, C.A., Rogers, R.R., and Monetta, A.M. (1993). Primitive dinosaur skeleton from Argentina and the early evolution of Dinosauria. *Nature* 361, 64–6.

Sereno, P.C. and Novas, F.E. (1992). The complete skull and skeleton of an early dinosaur. *Science*, 258, 1137–40.

Sewertzoff, A.N. (1931). Studien über die Reduktion der Organe der Wirbeltiere. *Zool Jahrb Abt Anat Ontogenie Tiere* 53, 611–99.

Shubin, N.H. (1994). The phylogeny of development and the origin of homology. In L. Grande and O. Rieppel (eds.), *Interpreting the Hierarchy of Nature*. San Diego: Academic Press.

Shubin, N.H. and Alberch, P. (1986). A morphogenetic approach to the origin and basic organization of the tetrapod limb. *Evol Biol* 20, 319–87.

Steiner, H. (1934). Über die embryonale Hand- und Fuss-Skelettanlage bei den Crocodiliern, sowie über ihre Beziehung zur Vogel-Flügelanlage und zur ursprünglichen Tetrapoden-Extremität. *Revue Suisse de Zoologie* 41, 383–96.

Steiner, H. and Anders, G. (1946). Zur Frage der Entstehung von Rudimenten. Die Reduktion der Gliedmassen von *Chalcides tridactylus* Laur. *Revue Suisse de Zoologie* 53, 537–46.

Swanson, A.B. and Brown, K.S. (1962). Hereditary triphalangeal thumb. *Journal of Heredity* 53, 259–65.

Vargas, A. and Fallon, J.F. (2005). Birds have dinosaur wings: the molecular evidence. *J Exp Zoolog B Mol Dev Evol* 304B, 86–90.

Vargas, A. O., Kohlsdorf, T., Fallon, J. F., VandenBrooks, J., and Wagner, G. P. (2008). The evolution of *HoxD-11* expression in the bird wing: insights from *Alligator mississippiensis*. PLo S ONE 3, e3325.

Wagner, G.P. and Gauthier, J.A. (1999). 1,2,3=2,3,4: a solution to the problem of the homology of the digits in the avian hand. *PNAS* 96, 5111–16.

Wagner, G.P. and Misof, B.Y. (1993). How can a character be developmentally constrained despite variation in developmental pathways? *J Evol Biol* 6, 449–55.

Warm, A., Di Pietro, C., D.'Agrosa, F., Gamblé, M., and Gaboardi, F. (1988). Non-opposable triphalangeal thumb in an Italian family. *J Med Genet* 25, 337–9.

Weiss, K.M. and Fullerton, S.M. (2000). Phenotypic drift and the evolution of genotype–phenotype relationships. *Theor Popul Biol* 57, 187–95.

Welscher, P. te., Zuniga, A., Kuijper, S., Drenth, T., Goedemans, H.J., Meijlink, F., and Zeller, R. (2002). Progression of vertebrate limb development through SHH-mediated counteraction of GLI3. *Science* 298, 827–30.

Welten, M.C.M., Verbeek, F.J., Meijer, A.H., and Richardson, M.K. (2005). Gene expression and digit homology in the chicken embryo wing. *Evolution & Development* 7, 18–28.

Zschabitz, A. (1998). Glycoconjugate expression and cartilage development of the cranial skeleton. *Acta Anat* 61, 254–74.

8

Functional analysis and character transformation

RICHARD A. RICHARDS

INTRODUCTION: PATTERN AND PROCESS

It is widely recognized that Darwin's theory of evolution incorporates, first, a historical "pattern" component asserting that species originated by divergent transformation from common ancestry, and that the similarities and differences among species reflect that pattern; and second, a "process" component asserting that this transformation occurs largely (but not exclusively) by the mechanism of natural selection. Given that these evolutionary patterns are products of evolutionary processes, it is unsurprising that systematists have traditionally relied on assumptions about process to reconstruct patterns of evolutionary transformation. But this use of process assumptions has been a source of dispute among systematists. Traditional "evolutionary systematists" have explicitly and systematically relied on process assumptions, while "cladists" have typically rejected the use of assumptions, on charges of circularity: if process assumptions are used to reconstruct evolutionary history, they can only be trivially confirmed by that reconstruction.

The primary focus of cladists' criticism has been functional analysis. The basic idea behind functional analysis is that once we understand how characters function, we can infer the operation of natural selection. And from that we can infer the likely directions of transformation, which can then be used to reconstruct the phylogenetic relationships of biological taxa. Cladists typically reject functional analysis, arguing that we should instead reconstruct evolutionary history on the basis of a

I would like to thank the editors and an anonymous referee for their support and suggestions. I am also in debt to Michael Ruse for the opportunity to develop and present the ideas in this paper.

theoretically neutral parsimony principle. It is not possible here to give a detailed account of the full range of the debate over functional analysis. It is possible, however, to get a sense of how the debate played out in a particular dispute – a dispute between two prominent figures, the evolutionary systematist Walter Bock, and the cladist Joel Cracraft. Bock argues that we can reconstruct the evolutionary past on the basis of character analysis, which begins with an analysis of form, then the observation of function and biological use. Bock justifies this approach on the assumption of the *transformational model* of change. Cracraft criticizes Bock's approach as misleading and liable to result in error. He bases his analysis on the assumption of the *taxic model* of change.

While I make no claims here about how well this debate is representative of the larger discussion, it is informative in that it reveals at least some of the issues in dispute. What is striking is that this apparently methodological argument is grounded to a significant degree in a disagreement about empirical and theoretical issues in evolutionary biology. And these empirical and theoretical commitments have implications relative to evidence, inferential strategies, and research programs. In the second section, I explain Walter Bock's framework for functional analysis; the third section presents Cracraft's objections to functional analysis; and the fourth section explains the justification Cracraft gives, based on the taxic model of macroevolution. Finally, in the concluding section, I explain how this particular dispute can be understood in light of, first, what counts as evidence; second, the preferred inferential strategy; and third, the relative merits of research programs based on form and function. What is striking is that while these sorts of issues are often seen as primarily methodological, for both Bock and Cracraft, they are ultimately empirical and theoretical issues.

FUNCTIONAL ANALYSIS

The dispute over functional analysis at issue here has its recent origins in the work of East German entomologist, Willi Hennig, who, in 1966, published an English translation of his 1950 book on systematics, *Grundzüge einer Theorie der Phylogenetischen Systematik*. This translation, titled *Phylogenetic Systematics*, provided the philosophical foundation for a revolution in biological systematics. The followers of Hennig, who came to be known as "cladists" (for their emphasis on the branching speciation associated with cladogenesis), challenged the then traditional approach to systematics, known as "evolutionary systematics" (or "evolutionary taxonomy") on two grounds – classificatory criteria and method of phylogenetic inference. Cladists argued first for an exclusively phylogenetic classification; and second, against the use of process assumptions to infer evolutionary patterns. Instead, they advocated a parsimony principle, which asserts that the best phylogenetic hypothesis is the one that requires the fewest assumptions of change. This parsimony principle, it was argued, is methodologically superior to other approaches because it assumed *only* that transformation had occurred – there were changes in character states, and nothing about the processes that produced this transformation.

Even though most systematists who work in the cladistic tradition (now commonly known as "phylogenetic systematics") do not focus on philosophical issues, there continues to be substantial opposition to the use of functional analysis in phylogenetic inference. Like many scientific disputes, the disagreement over functional analysis cannot be fully characterized in a simple and unequivocal manner. Individual systematists have varying philosophical and theoretical commitments that may change over time. Nonetheless, a symposium on functional analysis held at the 1979 Annual Meeting of the American Society of Zoologists reveals a prominent area of contention. The papers presented at this symposium were later published in a 1981 issue of the journal *American Zoologist*. At this 1979 symposium, and in the 1981 journal issue, Walter Bock represents the views of evolutionary systematics which advocates functional analysis, while cladist Joel Cracraft argues for its rejection.

According to Bock, there are two parts to biological systematics. First, is the formulation of phylogenetic hypotheses, which are then tested against the taxonomic properties of characters (Bock 1981: 5). Phylogenetic hypotheses might involve the postulation of a "sister grouping" among taxa, asserting that two or more taxa have a more recent common ancestor than other taxa under consideration. Or they might involve the postulation of an ancestor–descendant relationship, asserting that one taxon was the ancestor of another. In Bock's approach, these hypotheses are then evaluated on the basis of the hypothesized "taxonomic properties" of the various characters present in each taxon. By taxonomic properties, Bock means whether a character is a homology, homoplasy, plesiomorphy (symplesiomorphy), apomorphy, or synapomorphy, relative to two or more taxa (Bock 1981: 14). Homologies are shared characters that originated in a single common ancestor; plesiomorphies (or symplesiomorphies) are homologies that are ancestral to all the taxa under consideration; synapomorphies are more recent, "derived" homologies, and are shared only by those taxa with relatively recent common ancestry. Homoplasies, on the other hand, are shared characters that originated independently in multiple ancestors.

Two questions are relevant to the establishment of the taxonomic properties of shared characters. First, are the characters related by common descent – are they both products of a single origination event in a common ancestor? The forelimbs of birds and primates are believed to have a common origin and are therefore taken to be homologies. Alternatively, are the characters products of independent origination events in different ancestors? The wings of birds and butterflies are homoplasies because they are believed to have independent origins in different ancestors. Second, what is the phylogenetic origin (taxonomic level) of the homologous characters? Is the origin of a trait in a common ancestor to all the taxa under consideration? In other words, is it an ancestral homology – plesiomorphy (or symplesiomorphy)? Or is the origin in a more recent taxon, one not ancestral to all the taxa under consideration? Is it a derived homology – synapomorphy?

It is easy to see why Bock thinks that phylogenetic hypotheses can be tested against these taxonomic properties. This is because from the taxonomic properties of two or more characters, we can infer something about the relationship of the taxa in which they appear. Two taxa *must be* phylogenetically related if they share a true homology. After all, there could be a single origin of that trait in a common ancestor only if there were a common ancestor. And if two taxa share a synapomorphy or derived homology, by definition they must have relatively recent common ancestry. Alternatively, if the character is an ancestral homology, then there is common, but not necessarily recent, ancestry. On the other hand, if a shared character is a homoplasy, then relatively recent common ancestry is much less likely. There must have been sufficient time for the characters to develop independently in the various taxa. If birds' and butterflies' wings are homplasies, for instance, and originated independently in different ancestors, there must have been sufficient time since their divergence from a common ancestor for "wings" to develop in each lineage.

The second part of systematics, according to Bock, is the formulation of hypotheses about the taxonomic properties of characters, and the testing of these hypotheses "against the empirical observations of these features" (Bock 1981: 5). Bock claims that this second part – the character analysis – is the most important part of systematics because "it constitutes the objective testing with the systematic study and provides the basis for one's conviction in the validity of a particular classification and/or phylogeny" (Bock 1981: 5). It is here, Bock argues, that we find the real empirical testing of phylogenetic hypotheses. But we cannot just look at a shared character and determine its taxonomic properties – whether it is a homology or not – and its phylogenetic origin. According to Bock, the taxonomic properties of characters are to be inferred on the basis of functional adaptive analysis (hereafter just "functional analysis").

According to Bock, functional analysis involves both historical-narrative (H-N) explanations and nomological-deductive (N-D) explanations. Phylogenetic hypotheses and taxonomic property hypotheses are historical-narrative explanations, because they make assertions about the history of taxa and of their characters (Bock 1981: 11). These historical statements, according to Bock, can explain the various features of current taxa and characters as products of the conditions and causal processes that operated in the respective cases of evolutionary change (Bock 1981: 9). Nomological-deductive explanations, on the other hand, are the ahistorical laws governing causal processes, and can explain how and why a particular transformation occurred. (For our purposes here, I will refer to Bock's H-N explanations simply as "historical hypotheses," and N-D explanations as "process hypotheses.")

What is important in Bock's approach is that the *historical* phylogenetic and taxonomic property hypotheses are to be evaluated in light of *process* hypotheses. This is because, according to Bock, historical hypotheses implicitly assume something about the causal processes and laws that operated in the sequences that produced the observed characters and taxa. Historical hypotheses are better insofar as they rely on assumptions about process that are probable or true; and worse insofar as these assumptions are false or improbable. The relevant process assumptions include, at a minimum, the modification of characters and taxa by evolution. But, Bock argues, any accepted mechanism of change will be relevant as well (Bock 1981: 9). The basic idea here is that what we know about the mechanisms of evolutionary change can help us infer the most likely kinds of change, and the particular sequences and directions of change in the evolution of taxa and characters. According to Bock, these process assumptions are to be incorporated into functional analysis, which is aimed at understanding the functional properties of features and organisms. He distinguishes functional analysis from morphological analysis, and then cites his 1965 paper.

> Under functional analysis, I include all studies leading to the understanding of functional properties of features (function is used throughout in the sense of Bock and van Wahlert, 1965). This work is dependent on accurate, detailed morphological description of features which must be done with an understanding of the functional properties of those features: hence feedback must exist between morphological and functional analysis. (Bock 1981: 11–12)

In this 1965 paper, Bock and co-athor Gerd van Wahlert distinguish *form*, *function*, *faculty*, and *biological role*. *Form* is the appearance or configuration of a morphological or behavioral feature.

> The form of a feature is simply its appearance, configuration, and so forth ... In morphology, the form would be the shape of the structure. In behavior, it would be the configuration of the display, including the involved structures, their movements, intensity and so forth. (Bock and van Wahlert 1965: 272–3)

The form of a feature may be fixed, or it can change over time. A bone's form changes slowly, while a muscle's form can change rapidly.

The *function* of a feature is what it does:

> Its function is its action, or simply how the feature works – as stemming from the physical and chemical properties of the form; a feature may have several functions that operate simultaneously or at different times. (Bock and van Wahlert 1965: 296)

A tendon may have the function to contract, and an ossified tendon may have the function of not contracting. The skin of mammals has the functions of preventing the "passage" of materials such as fluids, dirt, and foreign organisms, and of reducing the flow of heat and stopping light rays. Bones function by resisting compression, tensile, and shearing stresses, as well as serving as a reservoir of mineral salts (Bock and van Wahlert 1965: 274–5). As this list suggests, a feature can have multiple functions, and functions may change if there is a change in form.

A *faculty* is a form–function complex: "The faculty, comprising a form and a function of the feature, is what the feature is capable of doing in the life of the organism" (Bock and van Wahlert 1965: 276). If a feature has one form and multiple functions, then it has multiple faculties – one associated with each form–function complex. So even if a bone were to have only one form, it could still have multiple faculties – one associated with being a reservoir of mineral salts, another associated with the resistance to compression, another with resistance to shearing stresses, and so on. And if a feature has multiple forms, it will have multiple faculties, even if it were to have only a single function (Bock and van Wahlert 1965: 276).

There may be a *biological role* associated with a feature. "The biological role of a faculty, and hence of the feature, may be defined as the use of the faculty by the organism in the course of its life history" (Bock and van Wahlert 1965: 278). A faculty (form–function complex) may be either utilized or non-utilized. Those faculties that get utilized acquire a biological role. A diving bird, for instance, may have a faculty for underwater "flying," but never use it. What is important here is that faculties that have biological roles associated are subject to natural selection on the basis of those roles:

> Each biological role of the faculty is under the influence of a set of selection forces, and consequently the number of different selection forces acting upon a faculty would depend upon the number of biological roles.
> (Bock and van Wahlert 1965: 279)

Adaptive analysis, then, takes what functional analysis reveals about form, function, faculty, and biological role of various features, and places it within a selectionist framework. The functioning of parts, and of the integrated whole, must produce an organism capable of surviving in the relevant environment. Postulated changes in form must maintain this survival ability:

> One must show that the suggested sequence of evolutionary changes of individual features in a phyletic lineage always results in functional organisms which are able to survive in the indicated environment.
> (Bock 1981: 12)

From the assumed operation of natural selection, then, we can draw inferences, not just about minimal survival ability, but also about the probability of various hypothetical character transformations, and the resulting character classifications. In other words, if we know something about the form and functioning of the parts and the integrated whole, and we know something about the biological role and the environment in which this functioning occurs, we can infer something about the likely directions and sequences of evolutionary change. And we can then determine whether a shared character is a homology, a homoplasy, and so on.

Bock identifies three main ways that functional-adaptive analysis works in character analysis and phylogenetic inference: first, to determine the probability that two characters are homologies; second, to determine the probable sequences of character transformation; and third, to determine the direction of transformation. He explains how this is supposed to work relative to four of his steps in character analysis (Bock 1981: 16–17):

1. Formulate and test hypotheses about homology.
2. Arrangement of homologues into transformation series.
3. Determine the polarity (direction of change) of a transformation series.
4. Establish the non-reversibility of change.

The first step in character analysis, to formulate and test hypotheses about homology, is not done directly on the basis of functional analysis. After all, homplasies (parallel and convergent evolution) often involve similar or identical functions. But functional analysis can nonetheless be used to evaluate the *probability* of hypotheses of homology or common origin.

> Confidence in the correctness of the test of individual homologues is judged by considering factors such as the complexity of the features, the relationship between form-functional properties of the feature and the selection forces governing its evolution (to ascertain the probability of independent origins, convergence, multiple evolutionary pathways, correlation between morphological and adaptive convergence, etc.). (Bock 1981, 16)

One of the ideas behind this use of functional and adaptive analysis is that we can identify those traits most likely to be the product of independent, convergent evolution, based on the fact that they are likely to be adaptations to specific environments. So if functional analysis suggests that a shared character is most likely convergent or the product of parallel adaptive evolution, then it is not likely to be a true homology that indicates common ancestry. For instance, because we know that the white coloration of foxes, rabbits, and bears functions in a particular way, and ultimately serves a specific biological role in snowy environments, therefore constituting an adaptation to that kind of environment, we can infer that the hypothesis of independent origins is more probable than the hypothesis of homology. We might formulate this "homology rule" as follows:

Homology rule: shared characters are less likely to be homologous if they are likely products of convergent or parallel adaptive evolution to a particular environment.

The value of this rule is obvious. If we can determine that some shared traits are *not* likely to be homologies, we can exclude them as evidence for common ancestry.

Once it has been determined which characters are most probably homologies, the second step is "the arrangement of homologues into transformation series, which reflect the evolution of the feature" (Bock 1981: 16). Bock argues that this is accomplished only on the basis of the known mechanisms of change.

> Because transformation series describe the presumed evolutionary histories of features, it is not possible to establish or to test them without a clear understanding of the mechanisms of adaptive evolutionary change of these features. And this understanding is not possible in the absence of thorough functional and adaptive analysis. (Bock 1981: 16)

Once the character is placed into a hypothetical transformation series, the direction of change – the "polarity" of the series – is to be determined. Here, functional adaptive analysis is based on the rule that change is always in the direction of the better-adapted state.

> Special mention must be made here of the use of postulated adaptive changes in the feature as a test of polarity of transformation series. This is based on the argument that evolutionary change will always be in the direction of the better adapted state. The direction of change can be argued strongly provided that a reasonable assessment of environmental forces can be made. (Bock 1981: 17)

This we might describe as a "polarity rule":

Polarity rule: in any character transformation, change will always be in the direction of the better-adapted state.

From this rule, we can determine whether a homology is ancestral (a symplesiomorphy) or derived (a synapomorphy). This is significant because if two contemporary taxa share a derived homology, they must, by definition, have a relatively recent common ancestor. But taxa that share an ancestral homology need not share a relatively recent common ancestor.

While Bock does not give an example of the polarity rule in his conference paper, examples appear in another conference paper authored by Wolfgang Gutmann (with the assistance of Bock). In this paper, Gutmann analyzes the hydrostatic "skeletons" of taxa within the coelomates. Within the taxa that constituted the coelomates, there are two forms of coelom – a metameric form and an oligomeric form. The metameric form is one in which "the coelom is subdivided into a large number of units by muscle bearing transverse septa." The oligomeric form, in contrast, "is one in which the body cavity is undivided or divided into two or three large cavities. Transverse septa are lacking …" (Gutmann 1981: 65). Gutmann (and Bock) asks which form is ancestral and which is derived:

> A major question in invertebrate phylogeny is whether the metameric coelom or the oligomeric coelom represents the primitive condition in the evolution of body cavities. Most workers … have postulated that the oligomeric condition is primitive because it has a simpler morphology in comparison to the morphologically more complex metameric condition.
> (Gutmann 1981: 65–6)

Gutmann questions the consensus view here, asking us to consider two possible transformation sequences. First, transformation could have occurred from the oligomeric form, with its few or single units and no transverse septa, to the metameric form, with its multiple units and muscle-bearing transverse septa. Or the transformation could have occurred from the metameric to the oligomeric. On the first transformation sequence, the transverse septa would develop and subdivide the coelom. On the second transformation sequence, the transverse septa would reduce and disappear (Gutmann 1981: 66).

Gutmann argues that functional-adaptive analysis can establish which of these transformation sequences is the most probable, and therefore which form is ancestral and which is derived. According to Gutmann, the transformation from the metameric to the oligomeric is plausible because each step preserves functioning:

> Evolution from the metameric to the oligomeric coelom involves the disappearance of the dissepiments and their transverse muscle. This can occur in gradual steps with the appearance of holes in the transverse septa by which fluid can move from one coelomic cavity to another. Enlargement of these holes would increase movement of fluid from one cavity to another and equalize pressure throughout the entire coelom. Evolutionary modification from a complete metameric condition to a fully oligomeric coelom can easily be conceived as a series of gradualistic steps as the change is a loss of existing features. (Gutmann 1981: 66)

But the second transformation, from the oligomeric to the metameric, is improbable because it requires passage through a dysfunctional state:

> Evolution from the oligomeric to the metameric coelom requires the formation of segmental transverse septa … Considering only the septa and transverse muscles, serious problems exist for a gradualistic explanation of their appearance in an oligomeric coelom and of their evolution to the metameric condition … The important function of the transverse septa is to subdivide the coelom into separate hydraulic chambers; this works only when the septa are complete. No functions and adaptive advantages have been suggested for incomplete septa and for the adaptive evolution from rudimentary septa to complete ones. (Gutmann 1981: 66)

The assumption here is that transverse septa only serve important functions when they are complete. And if development of the septa would require an incomplete stage, when there were only rudimentary septa that did not subdivide the coelom, then there is no reason to believe the septa could be formed by natural selection. Gutmann concludes that the first transformation sequence – from more complex metameric to the simpler oligomeric – is the correct one, because it is more probable in terms of the various functions and the operation of natural selection.

Using this analysis, Gutmann provides a hypothetical transformation sequence in the transformation of the deuterostome coelomates, from the primitive acoelomate form through to the derived notochord, with intermediate steps in the metameric and oligomeric coeloms (fig. 8.1).

A is the most ancestral state in this sequence, while F is the most derived. The cross sections indicate the metameric form in C and D, the oligomeric form in E, and the oligomeric form with notochord in F. There are four stages here, with three transformation events. First is the acoelomate stage, with a muscular grid surrounding a gel-like filling (A and B). The functioning in this state/form is movement for food uptake and some locomotion. Second is the metameric coelomate state, with true cavities, and the replacement of the gel with a fluid. This state provides for increased locomotion and the appearance of burrowing ability (C and D). The third state is the appearance of the oligomeric form, with fewer units, and greater flexibility, facilitating locomotion (E). Finally, is the appearance of a notochord (F), which, for a variety of reasons, results in greater efficiency of locomotion (Gutmann 1981: 67–9). In each of these transformation events, Gutman asserts an increase in functional efficiency and fitness.

Surely this approach, as worked out by Bock and illustrated by Gutmann, is plausible. Why should we not make inferences about character transformation, and the phylogenetic relationships of taxa on

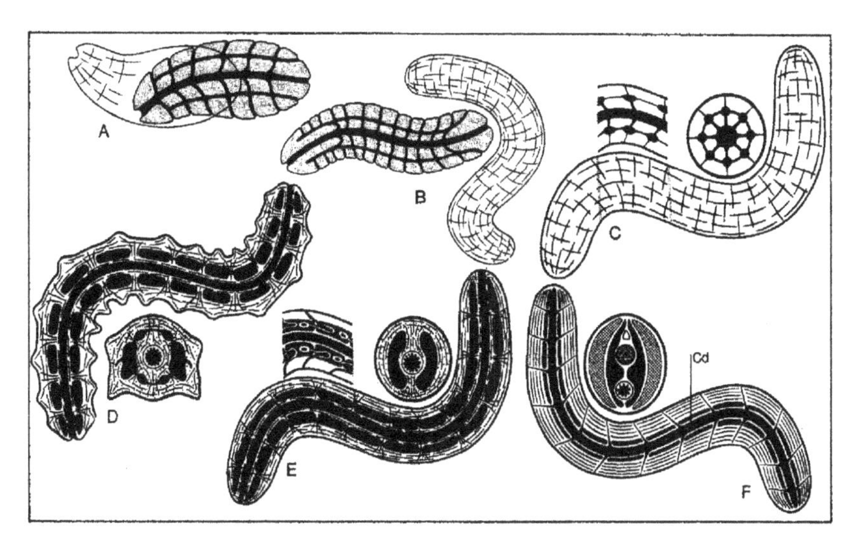

Figure 8.1 Model of hydraulic body construction in coelomates. The emergence of fluid-filled cavities in lower animals required the preceding existence of jelly-like support of the body frame, a situation well known from ctenophora. Jelly-like connective tissue was the mechanical condition for the functioning of muscles, and the precondition for the formation of canals by which digestion and distribution of food could be effected. A. Bending movements serving for improved food uptake into the canals could gradually change into locomotion by bending movements. B. The elongated worm shape (the stages are depicted in a shortened form) is mandatory for all possible locomotor activities because it compensates momenta generated by the movement of one body region by counteractions of other segments. C. Enlargement of some portions of the canal system provided fluid-filled spaces that reduced the rigidity of the jelly and resulted in the formation of fluid-filled hydraulic constructions. D. The muscular grid enforcing control of the elongated body shape permitted the emergence of many small fluid-filled cavities. Thus, metamerism of the coelomic cavities was the result of the biomechanical requirements of the system. E. The enlargement of canal portions in the lateral position offered maximal benefit for worms swimming by horizontal bendings because of the reduction of rigidity in the flanks which are subject to maximal deformations. The nephridial canals of metameric coelomates are explained as remains of the ancestral canal system. F. The chordate is characterized by a flexible, fluid-filled rod, the notochord (Cd), that keeps the body constant in length.

the basis of what we know about the processes that operate in evolutionary change? From the physiological, physical, chemical, and mechanical properties of characters, we can infer functioning. From this inferred functioning and observation of biological use, we can infer the operation of natural selection. And from the operation of natural selection, we can determine the likely paths of evolutionary change. Finally, from all this, we can infer the phylogenetic relationships of taxa. If the metameric coelom is ancestral to the oligomeric coelom, that tells us that the ancestral form is metameric, and the oligomeric is derived. If this is so, then the oligomeric coelom, at some phylogenetic level, is a derived homology (synapomorphy) and is evidence of a phylogenetic relationship.

THE CLADISTIC CHALLENGE: CRACRAFT AND WRIGHT'S RULE

In the 1979 conference on functional analysis and the 1981 journal issue devoted to that conference, Joel Cracraft represents a cladistic approach, and argues that functional-adaptive analysis is neither necessary nor useful in phylogenetic inference. He first argues that no process assumptions are required – except that there has been evolutionary transformation (Cracraft 1981: 24). Then he describes and rejects Bock's rules, arguing that there is no way to determine the relevant probabilities, transformation sequences, and directions of transformation (Cracraft 1981: 27, 29). He gives two reasons, one based on the availability of evidence, the other on the relevance of functional analysis.

Cracraft argues first that functional analysis requires knowledge about: (1) heritability; (2) genetic variance and its relationship to intrapopulational phenotypic variability; (3) the relationship between intrapopulational variation and variation in fitness; and (4) the nature of natural selection (Cracraft 1981; 31). The basic idea is that if character analysis is to be based on natural selection, we must be able to establish the factors relevant to the operation of natural selection. We need to know that the character is heritable – it has a genetic basis. We need to know something about the population in which it is found – the genetic and phenotypic variability within the population. Then we need to know the respective fitness values for the various phenotypes. Once we do this, then, perhaps, we can say something about the functioning of natural selection. We can presumably identify the fittest heritable phenotypes that would be most likely found in future populations, and therefore predict the makeup of future populations. If we know all this, we can make inferences about the probability of certain character transformations. The problem with functional analysis, according to Cracraft, is that this information is simply not available (Cracraft 1981: 31). Cracraft concludes that what evolutionary systematists are really doing in functional analysis is inventing functional and adaptation stories to support whichever phylogenetic hypotheses they prefer, on subjective or "authoritarian" grounds (Cracraft 1981: 27–9).

But it is not clear that Cracraft really disagrees with Bock. Bock is clear about the difficulties in getting the right kind of information in functional analysis.

> Study of the form and function of features in the laboratory, no matter how thoroughly done, is not a sufficient basis on which to determine their adaptive significance. It is essential to do field work on the biological role of the features, on the environmental demands on the organism and on the plausible selection forces. Only then, is it possible to determine relationship between the feature and the selection force, to judge whether an adaptation exists, and to estimate its degree of goodness. The total work required to ascertain adaptations is far more difficult than usually believed, and errors can be made in the establishment of an adaptation even when the analysis is complete and thorough. (Bock 1981: 12)

Bock concludes: "Far fewer features have been shown to be adaptations in any detail and with any assurance than believed by most biologists" (Bock 1981: 12). Although Bock then goes on to argue that there are at least some examples where the adaptiveness of a feature is well supported (such as industrial melanism in moths), his caution is striking. He seems to be agreeing with Cracraft that functional analysis requires information not easily or often available. The bottom line is that the availability of evidence does not appear to be the basis for the dispute.

Cracraft's second objection to functional analysis is its irrelevance. In the following passage, Cracraft is unqualified in his rejection.

> How do we go from functional data to probability levels? My interpretation is that the statement "the probability that evolution went from A to B is X" is *totally untestable*; in cases such as this, *there is no way* to evaluate the probability of a character transformation using functional data. (Cracraft 1981: 29, emphasis added)

Surely this is too strong – if the only issue were the availability of evidence for the operation of natural selection. Is it not possible to acquire *some* evidence that can tell us *something* about functioning and natural selection, and *something* about the probability of some transformations? Cracraft does not work out the details of this argument in this symposium paper, and we shall have to look elsewhere to fully understand it.

But he does make it clear here what is at issue, expressing doubt about the causal role of natural selection:

> Natural selection cannot be claimed to be a universal causal mechanism for the origin and maintenance of phenotypic diversity and design. This is not to say, of course, that natural selection is not operating or that it is unimportant in evolution, only that it cannot be applied as a general law in evolutionary explanation. (Cracraft 1981: 33)

The problem with functional adaptive analysis, according to Cracraft, is that it assumes the operation of natural selection beyond the intrapopulational level.

> Natural selection as a causal mechanism is widely accepted within evolutionary biology more for heuristic, common sense reasons than for any real understanding of its action in natural populations. Moreover, it is clear that statements about natural selection should be restricted to the intrapopulational level of analysis; extending natural selection to specific and supraspecific level … further confuses the issue of process analysis. (Cracraft 1981: 34–5)

It appears that Cracraft's real objection to functional analysis is what it assumes about the operation of natural selection in macroevolutionary change – change beyond the intrapopulational level.

Bock is seemingly aware of this as an issue, arguing that the value of functional adaptive analysis depends on the acceptance of the modern synthesis (Bock 1981: 8). And he self-consciously announces his assumption of the reductionist "synthetic" model of macroevolution: "[M]y analysis is based on the acceptance of the reductionist model of macroevolutionary change postulated under the synthetic theory of evolution" (Bock 1981: 11). For Bock, the synthetic theory of evolution is reductionist because it assumes macroevolutionary change to be nothing more than the sum of the changes within populations by natural selection. What is clear is that Bock and Cracraft see this disagreement about functional adaptive analysis to be, to some significant degree, a disagreement about macroevolutionary theory.

In the symposium papers, neither Bock nor Cracraft fully explores the significance of assumptions about macroevolutionary theory for the relevance of functional adaptive analysis. We can, however, turn to a book cited by Cracraft in this symposium paper, and co-authored by Cracraft with Niles Eldredge. *Phylogenetic Patterns and the Evolutionary Process* was published in 1980, after the symposium, but before the appearance of these papers in the journal issue. In this book, Eldredge and Cracraft quote Sewell Wright and cite "Wright's rule":

> Wright (1967: 120) wrote: "With respect to the long term aspects of the evolution of higher categories, the stochastic process is speciation. This was treated as *directed* above but may be essentially *random* with respect to the subsequent course of macro-evolution. The directing process here is selection between competing species often belonging to different higher categories." Eldredge and Gould (1972: 112, figure 5–10) utilized this notion as the basis of their reconciliation of the existence of long-term trends (i.e., net, directional changes in one or more morphological features within a monophyletic group through geologic time), with their rejection of the hypothesis that net, within-species variation is usually, or even commonly, directional through time. Stanley (1975) clarified and expanded this line of argument, referring to "species selection," and concluded that inter-specific evolutionary phenomena are in fact "decoupled," from intra-specific phenomena. Gould (1977b) and Gould and Eldredge (1977) used the expression "Wright's Rule" to refer to this aspect of the decoupling of phenomenological levels. (Eldredge and Cracraft 1980: 275)

There are two important ideas in this rather long, complicated passage: first, the "decoupling" of long-term, macroevolutionary trends from intra-specific, short-term, microevolutionary trends; and second, the causal account of this decoupling.

The first idea, the "decoupling of phenomenological levels," can be understood in terms of Wright's rule, which can be formulated as follows:

Wright's rule: macroevolutionary (long-term) trends are random relative to microevolutionary (short-term) trends.

What this implies is that short-term, microevolutionary trends can go in one direction, while long-term, macroevolutionary trends go in another. Wright's rule is illustrated in fig. 8.2 (Eldredge and Cracraft 1980: 285).

These four lineages a–d represent types of relationships between microevolutionary trends – change within each segment – and macroevolutionary trends – change over sequential segments. Eldredge and Cracraft describe the change in a, where macroevolutionary change is just the sum of microevolutionary change, as "consistent with the hypothesis that within-species variation and the direction of change are responsible for among-species patterns." In a, microevolution and macroevolution are congruent. Change in b "shows within-species patterns similar to the total among-species trend, likewise consistent with the above-stated hypothesis, but also consistent with the hypothesis that

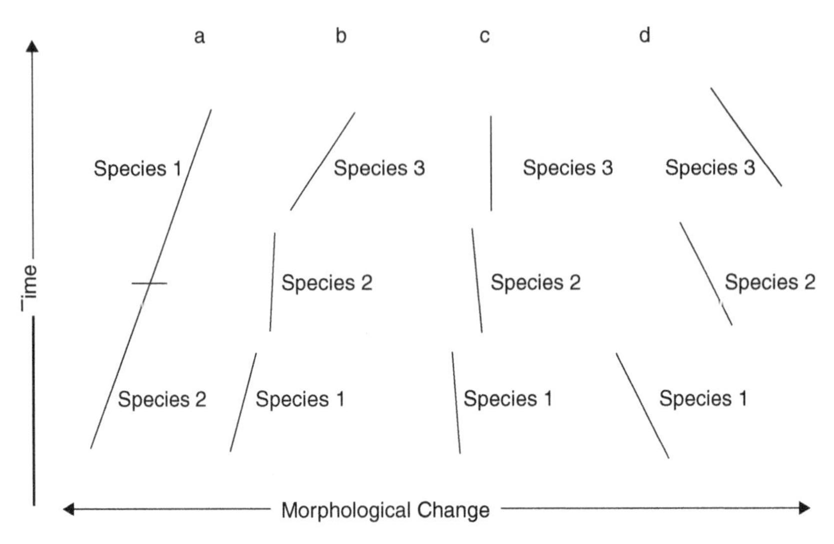

Figure 8.2 Relations between within- and among-species time-averaged patterns of variation. a. A hypothetical situation in which within-species direction of change through time is continuous and unbroken across species "boundaries." This type of change is consistent with the hypothesis that within-species variation and direction of change are responsible for among-species patterns. b. A situation in which the fossil record of three species shows within-species patterns similar to the total among-species trend, likewise consistent with the above-stated hypothesis, but also consistent with the hypothesis that the levels are decoupled. c, d. Directed, net change (a trend) among ancestors and descendants where the net change among species is inconsistent with the hypothesis that such trends are the direct product of within-species patterns of directional change.

the levels are decoupled." In c and d, we find "directed, net change (a trend) among ancestors and descendants where the net change among species is inconsistent with the hypothesis that such trends are the direct product of within-species patterns of directional change" (Eldredge and Cracraft 1980: 285). What is important here is that only in a is macro-evolutionary transformation a mere summation of microevolutionary transformation. And in c and d, microevolutionary and macroevolutionary trends are actually in opposition. If what happens in c and d occurs with some regularity, then Wright's rule – that macroevolutionary trends are random relative to microevolutionary trends – seems to be vindicated.

Some paleontologists now believe that the evolution of horses exemplifies this decoupling and randomness. In the evolutionary history of horses, ancient species were, on average, smaller than modern species; therefore, the macroevolutionary trend has been from small to large.

Horse evolution – from small size to large size – has traditionally been taken to be an instance of Cope's rule that lineages tend to evolve toward increasing size. This trend has also been taken to be a consequence of the greater fitness of large size; horses gradually became larger because natural selection favored increased size (Futuyma 1979: 139).

But it is now widely accepted that the evolution of horses has not been linear as this scenario might suggest. Rather, it seems to be more like a bush, with various branches leading in different directions, and virtually all branches becoming extinct. Most importantly, the trend toward increasing size is not found in all the individual branches. Larger size had been taken to be a general advantage, because, among other things, larger body size results in a reduction of potential predators. If being big were in fact a general advantage, and conferred a selective advantage on bigger individuals, then we should expect to find change within the branches (species) toward bigger morphotypes. But there is reason to believe, instead, that within the various branches of horse evolution, there has also been a significant trending toward decreasing size (MacFadden 1985: 255, 1986: 365. It just so happens that the lineages that survived are the ones that increased in size. If within the various lineages, size did not increase as a general rule, then horse evolution seems to have proceeded, not as it does in a or b, but as it does in c, or d. What is important here is that in c and d, and in horse evolution, species level change – microevolution – is not a reliable indicator of macroevolutionary change. David Jablonski, who has found a similar pattern of trends in mollusks, expresses doubt about the value of microevolutionary processes such as natural selection in explaining macroevolutionary trends.

> The selective forces often held to drive Cope's rule (for example, advantages of large size in defense, mating success, predatory ability and resistance to environmental extremes) generally imply the evolutionary loss of small-bodied forms. However, the fossil record shows that increases in the maximum body size within clades do not require such mechanisms, but can simply represent one limb of an expanding size range, and that directional decreases in size are no less frequent than increases. (Jablonski 1997: 252)

Jablonski concludes that "macroevolutionary patterns need not be simple extensions of those seen at the level of individual organisms over time" (Jablonski 1997: 252). (Some of the debate over Cope's rule has focused on "passive" explanations for directionality in trends that are not "driven" by a single "pervasive forcing field" such as natural selection, but by multiple factors (see McShea 1994: 1751).

If this account of horse evolution is correct, we can see why functional adaptive analysis would be problematic with regard to the polarity rule. Microevolutionary trends – trends within the various horse species – have gone both ways, toward increasing size and decreasing size. In either case, the assumption of functional analysis is that these trends within species have been a consequence of natural selection on the functioning of size. And if functional adaptive analysis were done *correctly*, it would reveal these trends. But there would be contradictory predictions about the polarity of the macroevolutionary trends. In the case of species that increased in size, functional adaptive analysis would predict that the transformation series of the horse lineages would have a polarity toward increasing size. But in the case of species that decreased in size, functional analysis would predict that the transformation series would have a polarity that indicated decreasing size. In this case, the polarity rule would be misleading. The point is this: even if functional adaptive analysis were done correctly, in the case of horse evolution it would not unambiguously and correctly indicate the polarity of transformation series relative to horse size.

THE TAXIC MODEL OF EVOLUTION

My intentions here are not to determine the validity of Wright's rule and the polarity rule, or the actual trends of horse evolution. Rather, my interests are in understanding the rejection of functional adaptive analysis by Cracraft (and perhaps other cladists). To fully understand Cracraft's rejection, we need to see why he appealed to Wright's rule. In the 1981 paper, he identifies two main reasons: the role of natural selection, and the decoupling due to speciation and species selection. First, he argued that "natural selection cannot be claimed to be a universal causal mechanism for the origin and maintenance of phenotypic diversity and design" (Cracraft 1981: 33). In particular, "natural selection should be restricted to the intrapopulational level of analysis" (Cracraft 1981: 34–5). Second, he claimed that the decoupling underlying Wright's rule was a consequence of speciation and species-level processes. The reasoning behind these two claims is found in the contrast Eldredge and Cracraft draw between the transformational and taxic models of macroevolution.

The transformation model of macroevolution is associated by Eldredge and Cracraft with "neo-Darwinism" (and the Modern Synthesis). The fundamental commitment of this model, according to Eldredge and Cracraft, is the central role of natural selection and adaptation.

> The transformational approach is generally developed in terms of adaptation, usually via natural selection. In this view, the central problem in evolution is the transformation of the genotype which occurs as organisms adapt in response to selection pressures. Adaptation is the ultimate *raison d'être* for change in gene frequency. (Eldredge and Cracraft 1980: 246)

On the transformational approach, adaptation by natural selection is taken to explain the patterns of evolutionary transformation at both the micro- and macroevolutionary levels, and the direction of level at the microevolutionary change gets extrapolated to the macroevolutionary level:

> The adaptational argument extends in detail to other aspects of macroevolutionary theory, involving directions of (morphological) evolutionary change as well as evolutionary rates. Perhaps the most classic, long-standing extrapolation of microevolutionary (within species) phenomena to among-species (macroevolutionary) phenomena involves directionality. In no other aspect of evolutionary theory is the confusion of within- and among-species phenomena most clearly seen. (Eldredge and Cracraft 1980: 266)

What is important here is that Eldredge and Cracraft take the transformational model of macroevolution to imply that the direction of long-term trends can be inferred on the basis of short-term trends. Since these short-term trends are determined by natural selection, we can infer them – and, ultimately, long-term trends – on the basis of what we can infer via functional analysis about the operation of natural selection. This is contrary, of course, to Wright's rule, which asserts that long-term trends are random relative to short-term trends.

Eldredge and Cracraft argue for the taxic model, and contrast it with the transformational model of macroevolution on the basis of three commitments. The first commitment is to what evolves – species rather than individual organisms and their properties (emphases added):

> Alongside the dominant theories of transformation has been another theme, that of the evolution of taxa. Although one rarely finds in the literature definitions of evolution conforming to this alternate approach, it is clear that many biologists ... who have studied species and theorized on speciation implicitly adopt the view that *evolution is quintessentially a matter of the origin of new taxa* – i.e. species. This, the "taxic" approach to evolution, simply alleges that at its core, *it is species that evolve – not individuals, and still less, anatomical or genetical properties of individuals.*
>
> (Eldredge and Cracraft 1980: 246, emphasis added)

What this implies, according to Eldredge and Cracraft, is that the important processes in terms of macroevolutionary change and trends are those processes governing the origin of new species and the differential survival of species. It is these processes that determine the macroevolutionary trends and the direction of character transformation.

The second commitment is to a theory of these processes. First is a theory of speciation that decouples the processes that operate within species, adaptation by natural selection, from the processes that operate among species in macroevolution. Speciation, according to Eldredge and Cracraft, is typically allopatric type II (the "peripheral isolates" model), and is a product of reproductive isolation. Whether a new species originates or not is a function of reproductive isolation, not adaptive change.

> Nowhere in contemporary works on speciation theory is the notion developed that speciation is fundamentally a product of adaptation. New adaptations, or the perfection of old ones, might be acquired, particularly in situations involving relatively small, peripherally isolated populations. But such adaptations, especially in allopatric situations, are incidental to the major phenomenon of the establishment of reproductive isolation. (Eldredge and Cracraft 1980: 270)

What is most important is that if speciation is a matter of reproductive isolation and not adaptive change, then factors like geographic isolation will be more important in determining whether a new species is formed than fitness or adaptive value. *It is the population that gets isolated which forms a new species, not the one that is most fit or adapted.* If this is so, we cannot predict that the new species will be of the most adapted, most fit form. In fact, the population that gets reproductively isolated might typically be *less fit* than other forms within the species. There would then be a decrease in fitness in the transformation sequence. This results in a "decoupling of micro and macroevolution" (Eldredge and Cracraft 1980: 277). Consequently, trends at one level (microevolutionary) cannot be extrapolated to trends at the other level (macroevolutionary) (Eldredge and Cracraft 1980: 326).

The second process commitment of the taxic model pertains to the survival of species. On the transformational approach, as characterized by the cladists, those species comprised of the best-adapted individual organisms are most likely to survive, because the best-adapted individual organisms are most likely to survive. The fitness of a species is a function of the fitness of its individual members, and so the probability of survival of a species is a function of the probability of the survival and reproduction of its individual members. Because of this connection, any adaptive trends within the species will naturally be expected in larger-scale trends as well. If larger size were advantageous to horses, then not only would we expect new horse species to be larger, but also that those that are larger would avoid extinction better. In contrast, in the taxic approach advocated by the cladists, the persistence of a species is no simple function of the fitness of its individuals. While it is true that the individuals comprising a surviving species must be minimally adapted, otherwise the species would become extinct, there are other factors – species selection and sorting processes – that are more important to differential species survival and long-term trends.

In the taxic approach, species selection is one "directing process" of macroevolution. In a passage cited above, Eldredge and Cracraft were explicit: "The directing process here is selection between competing species often belonging to different higher categories" (Eldredge and Cracraft 1980: 274–5). But they also argue that non-selection processes affect macroevolutionary trends.

> At the macroevolutionary level, interspecific trends may reflect some sort of "species selection," but could also arise, one might assume, from chance. The argument in the latter case mirrors that for genetic drift ... Similarly, while extinctions are clearly ... all caused by some particular concatenation of biotic and abiotic processes, the accidental ("bad luck") component of individual cases of species extinction is sufficiently apparent to prohibit the elaboration of a purely deterministic theory of macroevolution. To formulate such a theory as purely deterministic would be naïve.
> (Eldredge and Cracraft 1980: 277)

I cannot here address the existence and nature of species selection and sorting, but what is important is that Cracraft (and Eldredge) takes these processes to imply that long-term trends cannot be predicted from the operation of natural selection, because they are not based on natural selection. The bottom line is that, as understood by Eldredge and Cracraft, natural selection has a much more limited role on the taxic model than on the transformation model. Non-selective, random processes determine the origin of species, as well as the differential survival of species.

> At this juncture, we reiterate our acknowledgment of "random" factors; a new species may appear as the accidental by-product of change in the physical geography of the area ... A species might last longer than its sister in the next valley for purely accidental reasons, the converse of the

> observation that many extinctions are unlucky accidents. Entire ecosystems can be degraded relatively quickly ... In such cases, involving thousands of species in many unrelated clades, an environmental event (literally a catastrophe) occurs that is utterly accidental with respect to individual species adaptations. (Eldredge and Cracraft 1980: 308)

From all this, we can see how the taxic model might imply Wright's rule – that macroevolutionary trends are random relative to microevolutionary trends – and that functional analysis cannot be used to infer character transformations. The transformation that occurs in speciation cannot be assumed to produce a fitter form, because the less fit form might plausibly be found in the population that becomes reproductively isolated. And a transformation sequence might not be in the direction of the greatest fitness, also due to the random factors operating in differential species survival. The sheer bad luck involved in extinction events may favor less fit forms. Surely this would count against the polarity rule as employed by Bock? The bottom line is that the value and relevance of functional analysis to phylogenetic inference depends not just on the availability of evidence, but on the validity of Wright's rule about patterns of micro- and macroevolutionary change, and the underlying causal models assumed to be operating in macroevolutionary change.

CONCLUSION

There are three ways we can think about this dispute between evolutionary systematist Walter Bock and cladist Joel Cracraft. First, we can think about it as a dispute over what counts as evidence; second, we can think about it as a dispute about inferential strategies; and finally, we can think about it as a dispute over the research programs associated with form and function.

The function-based inference strategy and research program of Bock begins with morphological analysis, determining the form – shape, size, and configuration – of morphological (and behavioral) character traits. Two inference rules – the homology rule and the polarity rule – are based on functional analysis and are used to infer character transformation sequences, which, in turn, are used to classify characters as ancestral and derived homologies, homoplasies, and so on. These character classifications are then used to infer phylogenetic relationships. The inferential strategy based on these two inference rules is justified by Bock on the basis of his views about what counts as evidence for the reconstruction of evolutionary transformation. Bock's evidential stance accepts not just morphological similarity, but the processes that he incorporates into functional analysis, particularly those believed to be relevant to the operation of natural selection.

The form-based inference strategy and research program of Cracraft similarly begins with morphological analysis. But Cracraft rejects the homology and polarity rules, relying instead on a parsimony principle, which asserts that the best phylogenetic hypothesis is the one that requires the fewest assumptions of character change. The details of the application of this parsimony principle are beyond the scope of this paper, but what is important is the rejection of the two inference rules used by Bock to infer character transformation sequences, and the fact that Cracraft rejects these rules based on his views about the causal processes that operate in evolutionary change. There is then an empirical and theoretical justification behind his rejection of the function-based research program, and his resistance to an integration of function and form into a single inference strategy and research program. Cracraft's inferential strategy is therefore a product of his views about what can constitute evidence in the reconstruction of evolutionary transformations. Whether function really is irrelevant to phylogenetic inference, as argued by Cracraft, is an empirical issue, to be decided only by investigation into the causal processes that operate in evolution, in particular the role and significance of natural selection. Here, appropriate methodology is dependent on the empirical facts, which are themselves in dispute.

REFERENCES

Bock, W. (1981). Functional-adaptive analysis in evolutionary classification. *American Zoologist* 21, 5–20.

Bock, W. and van Wahlert, G. (1965). Adaptation and the form–function complex. *Evolution* 19(3), 269–99.

Cracraft, J. (1981). The use of functional and adaptive criteria in phylogenetic systematics. *American Zoologist* 21, 21–36.

Eldredge, N. and Cracraft, J. (1980). *Phylogenetic Patterns and the Evolutionary Process*. New York: Columbia University Press.

Futuyma, D. (1979). *Evolutionary Biology*. Sunderland, MA: Sinauer Associates Inc.

Gutmann, W.F. (1981). Relationships between invertebrate phyla based on functional mechanical analysis of the hydrostatic skeleton. *American Zoologist* 21, 63–81.

Hennig, W. (1966). *Phylogenetic Systematics*. Translated by D. Davis and R. Zangerl. Chicago: University of Illinois Press.

Jablonski, D. (1997). Body size evolution in cretaceous molluscs and the status of cope's rule. *Nature* 385, 250–2.

MacFadden, B. 1985. Patterns of phylogeny and rates of evolution in fossil horses: hipparions from the Miocene and Pliocene of North America. *Paleobiology* 11(3), 245–57.

(1986). Fossil horses from "Eohippus" (Hyracotherium) to Equus: scaling, Cope's law and the evolution of body size. *Paleobiology* 12(4), 355–69.

McShea, D. (1994). Mechanisms of large-scale evolutionary trends. *Evolution* 48(6), 1747–63.

9

The nature of constraints

ROGER SANSOM

INTRODUCTION

One of the most striking features of organisms is how well they are suited to live and reproduce in their environments. In general, we are impressed by the level of engineering necessary for an eye to see, or a bird to fly, or the potency of a snake's venom. And the closer one looks, the more marvels one finds, from the manifold adaptive capacities of the liver to the mechanisms of DNA proofreading. With his theory of evolution by natural selection, Charles Darwin provided the naturalistic explanation for such design that is now fundamental to the science of biology.

Although the theory of natural selection is considered Darwin's greatest contribution, his stated primary goal in publishing the *Origin of Species* was to argue that different species were of common descent.

> Although much remains obscure, and will long remain obscure, I can entertain no doubt, after the most deliberate study and dispassionate judgment of which I am capable, that the view which most naturalists entertain, and which I formerly entertained – namely, that each species has been independently created – is erroneous. I am fully convinced that species are not immutable; but that those belonging to what are called the same genera are lineal descendants of some other and generally extinct species, in the same manner as the acknowledged varieties of any one species are the descendants of that species. Furthermore, I am convinced that Natural Selection has been the main but not exclusive means of modification. (Darwin 1859: 6)

Darwin's evidence for common descent was "mutual affinities of organic beings, on their embryological relations, their geographical distribution, geological succession, and other such facts" (Darwin 1859: 3).

Generally, species that many consider to be quite different, such as mice and dolphins, have a great deal in common. These two species share a common general body plan of four limbs, a rib cage containing the vital organs, and so on. The closer one looks, the more similarities one finds, even between very different species. For example, all organisms share DNA (except for RNA viruses, which may not satisfy the criteria for being organisms). Given that organisms face different selective pressure, and assuming that the same solution is not typically optimal for all functional demands, these similarities show a restriction on the variation presented to natural selection. They are the result of constraints on evolution. Darwin is most remembered for his theory of evolution by natural selection, but he considered his most important contribution to be the argument for common descent, and his most important evidence of this was evidence of constraints.

Stephen Gould (1989) sees evolutionary change as the interplay between natural selection and developmental constraints. Much effort has gone into the analysis of natural selection, but far less has gone into the analysis of constraints, which is the subject of this paper (although, see Alberch 1980, 1982, 1989; Dullemeijer 1980; Gould 1989; Gould and Lewontin 1979; Maynard Smith *et al.* 1985; Resnik 1995; Sansom 2003; Schlosser 2007; Schwenk 1994, 2001; Wagner and Schwenk 2000; Wake 1982; Webster and Goodwin 1996; Whyte 1965).

SELECTION UNDER CONSTRAINT

The variation of all organisms is constrained (Maynard Smith *et al.* 1985). The simplest aspect of the relationship between developmental constraints and natural selection is to think of constraints biasing the variants available for selection.

Constraints deal in the metaphysically murky business of limits of possibility. While some philosophers might wonder about the metaphysical nature of possibility (e.g. Jacob 1992; Lewis 1986), it is good to see that scientists get into the business of defining the concept and investigating empirical examples. Maynard Smith *et al.* (1985) define a developmental constraint as "a bias on the production of variant phenotypes or a limitation on phenotypic variability caused by the structure, character, composition, or dynamics of the developmental system" (p. 266).

One of the simplest types of constraint is a correlation between two trait values. Frankino, Zwaan, Stern, and Brakefield (2005) accept

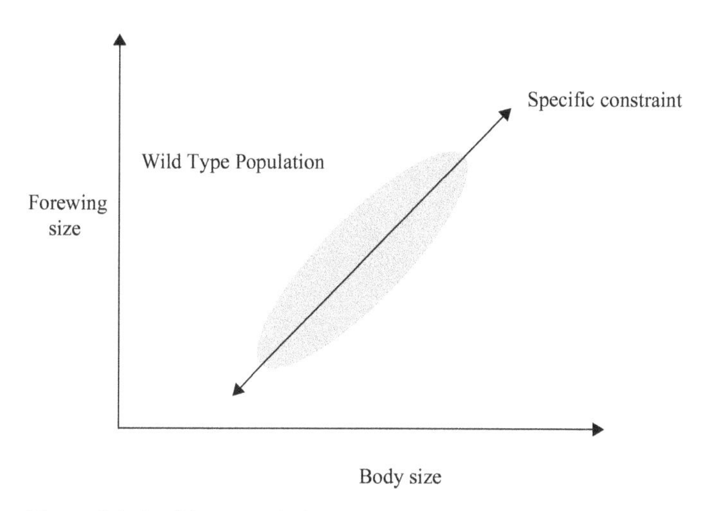

Figure 9.1 Positive correlation between body size and forewing size in wild type *B. anynana*

Maynard Smith *et al.*'s definition of constraints in presenting their investigation of the developmental constraints concerning the allometric relationship between forewing size and body size in the butterfly *Bicyclus anynana*. They provide an excellent empirical investigation of how the structure, character, composition, or dynamics of the developmental system bias the production of phenotypic variability (see also Brakefield, ch. 5 in this volume).

Brakefield measured the wild-type variation of combinations of two traits of all individuals in a population born into a particular generation and found that not all combinations were represented. Organisms with high body mass are biased toward having high forewing area. The combinations of body size and forewing size of each member of the population fall within the shaded area shown in fig. 9.1 (simplified from Frankino *et al.* 2005).

This example of the relationship between two quantitative character trait values is typical because most scale positively. Having noted this positive allometric relationship between body size and forewing size in the population, we would expect selection for higher forewing size to result in a population of higher body size too. In doing so, we induce that the population is under constraint represented by the line in fig. 9.1. After all, such a correlation between traits is an example of a limitation on phenotypic variability caused by the structure, character, composition,

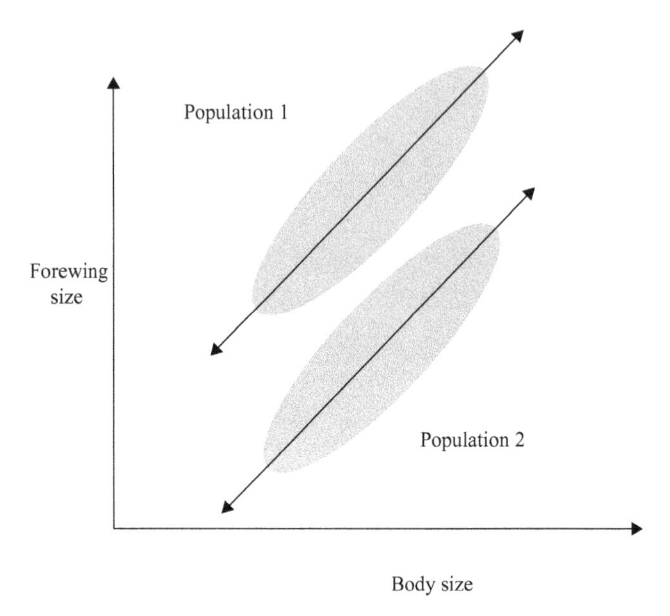

Figure 9.2. Positive correlation between body size and forewing size in *B. anynana* artificially selected for high and low wing-loading.

or dynamics of the developmental system, thereby satisfying the definition of constraints provided by Maynard Smith *et al.* (1985).

The ratio of wing area to body size is called wing-loading. It is directly related to flying ability (Vogel 2003). Flying ability is directly related to fitness, because flying is important in male competition and courtship of females (Joron and Brakefield 2003). Wing-loading can be thought of as a trait in itself. Brakefield and his colleagues investigated whether wing-loading can be changed with selection. They artificially selected for low wing-loading (a low ratio of wing area to body size) in Population 1, and high wing-loading in Population 2. Given that there is some variety in the ratio in the wild-type population, it is not surprising that they were able to produce divergence between wing-loading mean values in the two populations (fig. 9.2).

If we accept the constraint shown in fig. 9.1 to be a developmental constraint, then Brakefield has shown that it can easily be changed by selection. This is not surprising, given that this constraint – that is, wing-loading – varies across butterfly species (Vogel 2003).

Correlations between trait values are examples of constraints, but Brakefield has shown that this particular correlation can be changed.

This leads naturally to the question: how constraining are constraints on evolution?

Brakefield draws the conclusion that they are not much of a constraint after all: "it is not internal developmental constraints, but rather external natural selection that is the primary force shaping the short-term evolution of morphological allometries in insects" (Frankino *et al.* 2005: 720). Finding insect morphological allometries highly malleable by artificial selection, he apparently concluded that they are unconstrained.

Constraints can change as the ontogenetic mechanisms responsible for them change. They need not be inescapable. Does Brakefield's work show that insect morphology is sufficiently unconstrained that study of insect allometric relationships is relatively unimportant? I think that such a conclusion would be premature, because the way that this constraint changed was itself constrained.

The results shown in fig. 9.2 show that allometric relationships can be changed by selection. However, these results also show that the way these constraints changed can itself be constrained. In the case of *B. anynana*, the variation in wing-loading was due to changes in forewing area (resulting in populations moving up and down in fig. 9.2) rather than body size (which would result in populations moving left and right). This indicates a greater constraint on body size than forewing area.

For Brakefield, "The surprising bias in the morphological basis of how the allometry evolved suggests that development may strongly influence how individual traits respond to selection on their scaling relationships" (Frankino *et al.* 2005: 720). Brakefield interprets his results as showing that natural selection is the "primary" force in shaping short-term evolution of insect allometric relationships, but the way that natural selection changes them is due to relevant developmental constraints. I think that this is a natural conclusion to draw, but I remain cautious for the following reason.

If constraints resulted in selection moving the populations shown in fig. 9.2 up and down, rather than left and right, in what sense can selection be considered primary and constraints, presumably, considered secondary? I think that the best answer to this question is that selection caused the movement in a general direction (movement upward and/or leftward for population 1 in fig. 9.2, for example), but constraints determined the specific direction taken by the population. This is the typical metaphorical role given to developmental constraints as setting up the possibilities for natural selection to pick from. Because natural selection is the provider of movement of a population through morphospace, it is considered in some sense privileged. I am cautious about privileging natural selection, however, because both natural selection and constraints determine the outcome of such an evolutionary transition.

While some constraints may easily be escaped by natural selection, the way this happens is itself constrained. Faced with this complexity, one could decide that only constraints that are sufficiently binding should count as genuine constraints. On this type of interpretation, Brakefield's results suggest that body size is constrained, but wing-load is not. In contrast to such an interpretation, Maynard Smith *et al.* (1985) are willing to take an inclusive approach, accepting that constraints lie on a continuum of "bindingness" (p. 267). Accepting such a continuum of constraints leads to the following view. Constraints are correlations between trait values in any generation before there is selection on that generation. A constraint may also be considered a trait value. Such a constraining trait value may itself be constrained by another developmental constraint. In this view, different populations are under at least somewhat different constraints. If natural selection is involved in evolutionary transitions that result in populations being under different constraints, then we need to face the possibility that natural selection selects constraints. In the next section, I will use another example of insect allometry to discuss constraint selection.

CONSTRAINT SELECTION

Nijhout and Emlen (1998) discovered that there is an inverse relationship in *Onthophagus acuminatus* (a dung beetle) between size of horn and size of eyes (each relative to body size). They show that large-horned and -eyed *O. acuminatus* are unlikely to become available to natural selection due to a limit on the number of variants of *O. acuminatus* that can develop.

This constraint may be adaptive. Broadly speaking, the male beetles have evolved two mating strategies open to them (Emlen 1997; Moczek and Emlen 2000). The first strategy is to stand guard at the mouth of a female's burrow to ensure that they are the only one to mate with her. Having large horns is adaptive for this strategy because it contributes to success in male combat (Moczek and Emlen 2000). The second strategy is to gain access to females by stealth, either by sneaking past a guard or actually digging an intercepting burrow (Emlen 1997). Having small horns and large eyes is adaptive for the second strategy.

While some adaptive trait values may be unreachable due to constraints, the example of this constraint on *O. acuminatus* shows that constraints can be adaptive if they limit trait values adaptively. It should be acknowledged that a constraint may now be adaptive because other trait values have evolved to make it so. For example, *O. acuminatus* may have evolved dimorphic behavioral strategies of guarding or sneaking into burrows to take advantage of the trait values that result from this constraint. Nevertheless, the adaptivity of this constraint raises an important point about the relationship between constraints and natural selection. If there are a variety of constraints within a population, and a constraint can be adaptive, then adaptive constraints can be selected. In this section, I shall investigate the nature of such selection.

One outcome of the effort that has gone into understanding natural selection is the units of selection debate. Richard Dawkins (1976) argued that the gene is the only unit of selection, because only genes are replicated across generations. Stephen J. Gould and David Hull together rejected gene selectionism. For Gould, units of selection "must be actors within the guts of the mechanism, not items in the calculus of results" (Gould 2001: 213). For Hull, each unit of selection must be an "interactor: an entity that directly interacts as a cohesive whole with its environment in such a way that replication is differential" (Hull 1980: 318). From this debate emerged the view that there are multiple units of selection expressing different types of traits.

The unit of selection for a trait is the smallest individual that directly expresses variety in that trait value to natural selection. There is gene selection, but it is limited to phenomena where the genes themselves are directly involved, such as selection for meiotic drive. Trait values are typically expressed at the level of individual organisms. One of the most controversial units of selection is groups of organisms, which some claim can be selected for group traits such as altruism between individuals (e.g. Sober and Wilson 1998). This unit has multiple organisms, which can express traits like cooperation between individuals.

Constraints determine the range of trait values that are possible. A single organism has only one combination of trait values, rather than a range. Therefore, a constraint can only be expressed by a unit of selection made up of multiple organisms.

Consider the lineage of organisms that descends from the first organism to have a particular constraint. If reproduction is sexual, then each generation might include descendants that share the constraint. A lineage of organisms that share the constraint is a constrained lineage. Each organism in a constrained lineage might share the same combination of trait values, but might also have a variety of constrained trait values, due to variety in the population entering the lineage by way of sexual reproduction. They do share the same constraint, and it determines their range of combinations of trait values. Therefore, I propose that selection that favors the constrained lineage is selection for the constraint. This is a unit of selection made up of multiple organisms in the same generation that I call a *legacy* (Sansom 2007).

Returning to our example, some ancestor of today's *O. acuminatus* was the first to have this constraint, due to some heritable mutation. Any selective advantage it had would have been due to its adaptive trait values. It may have been the first to have horns on its head and also had smaller eyes than the rest of the population. Its descendants that shared the constraint made up multiple legacies that had different combinations of trait values (some had large horns and small eyes, while others had small horns and large eyes), due to variation in the population. Selection that favored these legacies was selection for their constraint.

Evolvability is defined as "the ability of random mutations to sometimes produce improvement" (Wagner and Altenberg 1996: 967). Constraints might not be selected only for compatibility with diversity in the population, but also for compatibility with novel mutations as they arise (Sansom 2007). Any such selection for constraints would take place in longer lineages, because it takes longer for variety to enter a lineage by way of mutation than sexual reproduction.

THE HIERARCHY OF CONSTRAINTS

Nijhout and Emlen (1998) investigate the mechanism that is responsible for the negative correlation between horn and eye size in *O. acuminatus*. They find that it is due to the fact that these features develop from the same limited pool of resources. Resources are limited because they are developed after the larvae have stopped eating. The negative correlation between these two features is due to the fact that they develop in close proximity to each other.

Understanding the mechanisms responsible for constraints helps us to understand how general a constraint is, and vice versa. Nijhout and Emlen find that negative correlations between values of traits that develop proximately are widespread in insects. For example, removal of hindwing imaginal discs from male caterpillars of *Precis coenia* during the second day of the final larval instar increased the relative dry weights of the adult thorax, but not its head and abdomen.

Maynard Smith *et al.* (1985) noted that constraints differ in their scope. The constraint between horn and eye size is of narrow scope. We would expect scarab beetles with thoracic horns to not have this constraint. In contrast, the constraint of negative correlation between traits that develop in proximity during metamorphosis constraint may apply to all insects that go through complete metamorphosis. Scarab beetles with thoracic horns do have a negative correlation between horn size and wing size (Kawano 1995).

Acknowledging the difference in scope between these two constraints insufficiently describes the relationship between them. A full description will also describe the fact that any organism that falls under Nijhout and Emlen's general constraint will fall under multiple, more specific constraints concerning particular trait values that are developed in close proximity. In addition, a mutation that alters one of these specific constraints will most likely result in a new similar constraint. For example, if a mutation were to move the horn of *O. acuminatus* from its head to its thorax, while the negative correlation between horn and eye size would be gone, we would expect a new negative correlation between horn and wing size to arise.

In an attempt to capture the complexities of the relationship between constraints that rely on the same developmental mechanisms, I propose that such constraints form a modal hierarchy. More specific constraints are lower in the hierarchy. For example, the constraint between horn and eye size in *O. acuminatus* will lie on the same level as the constraint between hindwing and thorax size in *P. coenia*. Governing both (and a great many other similar constraints) in the hierarchy is the general constraint about the inverse relationship between proximately developing traits in holometabolous insects.

Higher-level constraints can never apply to fewer organisms than lower-level constraints that they govern, and will nearly always apply to more. However, with this greater scope comes less specificity. To know that trait values that develop in close proximity have inverse relationships is not to know which trait values develop in close proximity. That added information is included at lower levels in the hierarchy. At the very top of the constraint hierarchy is one that governs all constraints. It may be so general as to be hard to describe in an informative way. Karl Niklas begins his investigation of form and function in this volume (ch. 3) with a morphospace limited by mathematical conceivability, which seems a likely candidate for the most general constraint. It governs all organisms but tells us very little about any. At the very bottom may be the constrained trait values of each organism, which strongly indicate the variety of progeny that it can reproduce.

Different constraints, located in different parts of a constraint hierarchy, and explained by different mechanisms, may contribute to the same trait value. For example, Brakefield's results suggest that there are two quite independent influences on wing-load. There is the positive correlation between wing size, and body size and also an independent constraint on body size.

This hierarchy of constraints is nothing like a phylogeny, which represents actual historical events. Rather, it represents modal structure. Governing is a primitive modal notion, and discovering governing relationships between constraints is an attempt to peer into the murky business of limits of possibility, alluded to at the beginning of the second section, and to represent what is possible and why.

CONCLUSION

The role of constraints in evolution is pervasive – all organisms are under many constraints, which determine the limits of possibility of organisms. Describing these possibilities represents theoretical and empirical difficulties. Despite this, there is growing interest in constraints, and theoretical and empirical work on them is accelerating. Evo Devo is a fast-growing area of biology and includes investigations of constraints and their evolution.

Given the importance of natural selection to the process of evolution, inevitably our understanding of constraints will be, at least in part, in terms of how they relate to natural selection. That relationship is complex. There is a sense in which they set the stage for natural selection and drift to move populations through morphospace. Lineages of different constraints are also available to natural selection for producing different ranges of trait values, given variety in sexually reproducing populations and mutations.

Our theoretical and empirical investigation of constraints and their role in evolution remains in its infancy and may develop to have a profound influence on our understanding of life.

REFERENCES

Alberch, P. (1980). Ontogenesis and morphological diversification. *American Zoologist* 20, 653–67.

(1982). The generative and regulatory roles of development in evolution. In D. Mossakowski and G. Roth (eds.), *Environmental Adaptation and Evolution*. Stuttgart: Fischer, pp. 19–36.

(1989). The logic of monsters: evidence for internal constraint in development and evolution. *Geobios. Mémoire spécial* 12, 21–57.

Darwin, C. (1859). *On the Origin of Species by Means of Natural Selection or the Preservation of Favoured Races in the Struggle for Life*. London: John Murray.

Dawkins, R. (1976). *The Selfish Gene*. Oxford University Press.

Dullemeijer, P. (1980). Functional morphology and evolutionary biology. *Acta Biotheoretica* 29, 151–250.

Emlen, D.J. (1997). Alternative mating tactics and male dimorphism in the horned beetle Onthophagus acuminatus (Coleoptera: Scarabaeidae). *Behav Ecol Sociobiol* 41, 335–41.

Frankino, W.A., Zwaan, B.J., Stern, D.L., and Brakefield, P.M. (2005). Natural selection and developmental constraints in the evolution of allometries. *Science* 4(307), 718–20.

Gould, S. (1989). A developmental constraint in cerion, with comments on the definition and interpretation of constraint in evolution. *Evolution* 43(3), 516–39.

(2001). The evolutionary definition of selective agency, validation of the theory of hierarchical selection, and fallacy of selfish gene. In R. Singh, C. Krimbas, D. Paul, and J. Beatty (eds.), *Thinking About Evolution*. Cambridge University Press, pp. 208–34.

Gould, S. and Lewontin, R. (1979). The spandrels of San Marco and the Panglossian paradigm: a critique of the adaptationist programme. *Proceedings of the Royal Society of London B* 205, 581–9.

Hull, D. (1980). Individuality and selection. *Annual Review of Ecology and Systematics* 11, 311–32.

Jacob, F. (1992). *The Possible and the Actual*. New York: Oxford University Press.

Joron, M. and Brakefield, P. (2003). Captivity masks inbreeding effects on male mating success in butterflies. *Nature* 424, 191–4.

Kawano, K. (1995). Horn and wing allometry and male dimorphism in giant rhinoceros beetles (Coleoptera: Scarabaeidae) of tropical Asia and America. *Annals of the Entomological Society of America* 88, 92–9.

Lewis, D. (1986). *On the Plurality of Worlds*. Oxford: Blackwell.

Maynard Smith, J., Burian, R., Kauffman, S., Alberch, P., Campbell, J., Goodwin, B., Lands, R., Raup, D., and Wolpert, L. (1985). Developmental constraints and evolution. *The Quarterly Review of Biology* 60, 265–87.

Moczek, A.P. and Emlen, D.J. (2000). Male horn dimorphism in the scarab beetle Onthophagus Taurus: do alternative reproductive tactics favour alternative phenotypes? *Anim Behav* 59, 459–66.

Nijhout, H.F. and Emlen, D.J. (1998). Competition among body parts in the development and evolution of insect morphology. *Proceedings of the National Academy of Sciences USA* 95, 3685–9.

Resnik, D. (1995). Developmental constraints and patterns: some pertinent distinctions. *Journal of Theoretical Biology* 173, 231–40.

Sansom, R. (2003). Constraining the adaptationism debate. *Biology and Philosophy* 18(4), 493–512(20).

(2007). Legacies of adaptive development. In R. Sansom and R. Brandon (eds.), *Integrating Evolution and Development: From Theory to Practice.* Cambridge, MA: MIT Press, pp. 173–94.

Schlosser, G. (2007). Functional and developmental constraints on life cycle evolution: an attempt on the architecture of constraints. In R. Sansom and R. Brandon (eds.), *Integrating Evolution and Development: From Theory to Practice*. Cambridge, MA: MIT Press, pp. 113–72.

Schwenk, K. (1994). A utilitarian approach to evolutionary constraint. *Zoology* 98, 251–62.

(2001). Functional units and their evolution. In G. Wagner (ed.), *The Character Concept in Evolutionary Biology*. San Diego: Academic Press, pp. 165–98.

Sober, E. and Wilson, D. S. (1998). *Unto Others: The Evolution and Psychology of Unselfish Behavior*. Cambridge, MA: Harvard University Press.

Vogel, S. (2003). *Comparative Biomechanics*. New Jersey: Princeton University Press.

Wagner, G. and Altenberg, L. (1996). Complex adaptations and the evolution of evolvability. *Evolution* 50(3), 967–76.

Wagner, G. and Schwenk, K. (2000). Evolutionarily stable configurations: functional integration and the evolution of phenotype stability. *Evolutionary Biology* 31, 155–217.

Wake, M.H. (1982). Diversity within a framework of constraints. Amphibian reproductive modes. In D. Mossakowski and G. Roth (eds.), *Environmental Adaptation and Evolution*. Stuttgart and New York: Gustav Fischer, pp. 87–106.

Webster, G. and Goodwin, B. (1996). *Form and Transformation*. Cambridge University Press.

Whyte, L. (1965). *Internal Factors in Evolution*. London: Tavistock Publications.

10

Toward a mechanistic Evo Devo

ANDREW L. HAMILTON

> *I have acquired the conviction that our biological theories must remain inadequate so long as we confine ourselves to the study of cells and persons and leave the psychologists, sociologists, and metaphysicians to deal with complex organisms.*
>
> W.M. Wheeler (1911)

INTRODUCTION: TWO DISTINCTIONS

Manfred Laubichler's contribution to this volume contains a subtle argument to the conclusion that in evolutionary developmental biology (Evo Devo), form and function are both best understood as kinds of *causes* at different scales that can be brought together in mechanistic explanations of phenotypic evolution. This argument is evocative not only because it leans heavily on mechanistic explanation – about which much more below – but also because there have been long periods during which form was not understood to have a causal aspect. When the phenotype is a passive by-product of evolution, the form does not do much work.

Evo Devo, of course, is partly built around denying the correctness of one-sided explanatory strategies, and thus, if Laubichler is correct, finds itself brushing up against another familiar distinction. When Ernst Mayr (1961) made his cut between proximate and ultimate causation, he recognized two sorts of causes in biology, but minced no words about which he took to be the more important. It is possible, of course, to agree with Mayr's distinction and disagree with him about which kind of causes matter most, just as it is possible to reject the nature–nurture dichotomy and still wonder whether nature or nurture is more important. The message here is not that Laubichler's reading of Evo Devo cannot be correct because it pushes us in the direction of distinctions that some find otiose. On the contrary: in noticing Evo Devo's attempt to bring together two kinds of causes that have often been held apart, Laubichler's epistemology of Evo Devo becomes a cautionary tale that raises an interesting set of questions. As Alessandro Minelli (2003) has pointed out, we have yet to hear much from the evolutionary side of the Evo Devo duo. When we do, what will happen? What reason is there to think that evolutionary biology and developmental biology will not re-fragment along ultimate-proximate lines?

Laubichler sees mechanistic explanation as a locus of integration for form and function, as well as for evolutionary biology and developmental biology. Here I will discuss a pair of issues that this approach raises. The first is not organism- or group-specific, but rather has to do with mechanistic explanation as applied to Evo Devo. In particular, I will try to deflate a concern over whether the recent focus on gene regulatory networks is a throwback to the thoroughgoing molecular reductionism of just a few years back, by pointing to some features of mechanistic explanation that have been drawn out by philosophers. This is a problem that I think I can dissolve.

Another issue that Laubichler's framing raises indirectly is specific to social insects, which are potentially an important study group for Evo Devo: if tightly integrated colonies of social insects are higher-level entities – "superorganisms" of some sort – and we are seeking mechanistic explanations, we have some work to do in articulating and testing accounts of how colony development connects with the accounts we have of the development of the organisms that constitute them. In other words, a levels-of-development problem falls out of combining mechanistic thinking with conceiving of colonies as higher-level units.

These issues bear on form and function in several ways, but the connections are oblique enough that some spelling out will be helpful. The conversation about reductionism in the next section concerns a source of worry about what shape explanations of form and function by way of a mechanistic Evo Devo take. If they are reductionistic in a certain sense, then many will want to resist them, or at least to call for something wider in scope as well. The challenge of framing an explanation for colony development examines a case of Evo Devo *in vivo*, as it were, and points to the project of integrating form with function by way of articulating developmental mechanisms at more than one level of organization, as well as the connections between them.

It is too early to tell, of course, whether the focus on mechanistic explanation will truly be a source of integration or will prevent a new schism along the old form–function boundary. My aim here is not to make the case for mechanistic explanation, but to defend it against a charge of which it is innocent, and to test it by pressing a hard but potentially important case. I hope that the latter initiates a conversation about what the assumptions, goals, and methods of research projects in Evo Devo stand to gain or lose from treating explanation as a bringing-together of form-related and function-related causes in a single mechanistic model.

MECHANISMS AND LEVELS OF ORGANIZATION

The integration of evolution and development that Laubichler envisions for Evo Devo is driven in large part by advances in detecting and understanding gene regulatory networks of the kind described by Eric Davidson and Douglas Erwin (2006). The idea is that gene regulatory networks tell us how changes in the regulation of gene expression issue in changes in the patterns and processes connected with growth, differentiation, and morphogenesis, which leads in turn to an account of evolution and development that is integrated by attention to the mechanisms for phenotypic variation.

One obvious concern about this understanding of the project of twenty-first-century Evo Devo is that it places the molecular gene at the very center of the explanatory framework. This has the potential to be worrying for two related reasons. First, the role of gene regulatory networks in this story will strike some as employing reductionistic thinking of the worst kind. What is so new and exciting, after all, about accounting for phenotypes by paying attention to how genes regulate development? Is this not just another attempt to explain organisms at the molecular level, with the difference that in the new account genes are a means to explain, rather than ignore, development? Second – and this is a closely related worry – critics of gene-centric biology have argued forcefully that privileging the molecular view fundamentally misconstrues the importance of the gene in explanations of phenotypes and behavior, partly because focusing on genes has meant failing to investigate many other potentially important causes for biological outcomes (Lewontin 2000). Development, of course, is among these other causes.

These objections can partly be answered by noticing that mechanistic explanations are reductionistic, but they need not lead to gene-centrism, and they need not lead anyone to ignore the overall system in favor of the actions of one "fundamental" part of it. They do not, that is, emphasize function at the expense of form (or vice versa). Consider the following conception of mechanism:

> A mechanism is a structure performing a function in virtue of its component parts, component operations, and their organization. The orchestrated functioning of the mechanism is responsible for one or more phenomena. (Bechtel and Abrahamsen 2005)

In this way of understanding mechanisms – which is different in nuanced ways from some of the other mechanism concepts in the philosophical literature, but is not by any means atypical – both form and function are central considerations (see also Glennan 2002; Machamer *et al.* 2000; Wimsatt 1976). Mechanistic explanation works by decomposing systems responsible for certain phenomena into their component (lower-level) parts and operations, and so is reductive (Bechtel and Hamilton 2007; Bechtel and Richardson 1993).

Mechanistic explanation is also inherently inter-level. It builds bridges between levels of organization by noting that the behavior of component parts is often different than the behavior of the system of interest (Bechtel and Hamilton 2007). Decomposition is a primary strategy of mechanistic approaches partly because it recognizes important subsystems at various levels of organization within the system of interest, and it is understood that the functioning of each of these subsystems depends on the outputs of other subsystems. By this way of thinking, no level of organization is fundamental, because each subsystem contributes importantly to the behavior of the target system, and no subsystem is assumed to behave in the same manner as the overall system. Another way to put this is that mechanistic thinking of this kind brings form together with function, by realizing that mere aggregation of parts does not account for system-level properties: the *organization* of the parts matters (Wimsatt 1974).

This summary of the basic commitments of mechanistic explanation relies heavily on the work of philosophers. I take it that I have shown how one might offer an explanation that is mechanistic without being naïvely reductionistic, overly myopic, and focused on one level of organization at the expense of others. Do practitioners of Evo Devo actually work this way? The answer to this question, as the papers in this volume and others show, is that they do. It is certainly the case, as Raff and Raff point out in chapter 4 of this volume, that "studies of evo-devo have until recently largely focused on understanding gene regulatory changes at the macroevolutionary scale." It is also the case that despite a focus on genes, much of this work has been couched in an inter-level framework. Among the basic concepts employed in this work are modularity, hierarchy, and emergence. Emergence requires paying attention to at least two levels, as does hierarchy. Modularity is often an explicitly inter-level notion because modules are frequently understood as component mechanisms with a larger unit (Raff and Sly 2000; Wagner 1996). In biology, looking for mechanisms is not always looking for genes, and explanation by a kind of reduction that keeps the whole system in view need not be unpalatable.

DEVELOPMENT AND LEVELS OF ORGANIZATION: THE CASE OF SOCIAL INSECTS

Having defended a mechanistic approach to Evo Devo from one source of criticism, I would now like to outline a challenge for this approach that arises from thinking about social insects as model systems for Evo Devo. The challenge is to connect developmental thinking with evolutionary thinking about behavior in social insects. Social insects are an interesting case because some of them form colonies that can be said to develop. My goal in this section is to point to interesting conceptual and empirical work to be undertaken in discovering the mechanisms for colony-level development and how they are related to organismal development. In order to see what the missing piece is that needs to be filled in, a rough survey of the field will be helpful.

The study of social insect behavior in historical context

Social insects pose a special problem for explanation because they engage in behaviors that appear to benefit others at fitness costs to themselves. Highly social honey bees, for instance, are descended from less social bees, and ultimately from much less social wasps, yet they live in colonies where generally only one female of a given generation reproduces. Where fitness is understood as an expected number of offspring, the explanatory challenge is to account for the evolution of a colony structure in which the direct fitness of all but a few organisms out of tens of thousands is effectively zero.

Three general kinds of solution are available, though none has achieved anything like consensus and all have important puzzles that remain to be solved. The first solution is Darwin's, from *The Descent of Man* (1871): make the group rather than the organism the unit of selection. A second, and often closely related strategy, is to understand the group as an organism. A third strategy is to account for social behaviors in terms of genetic relatedness. On this view, helping one's sibling to reproduce is to contribute to the passing on of the shared part of one's own genetic complement, or as W.D. Hamilton famously put it:

> To express the matter more vividly, in the world of our model organisms, whose behavior is determined strictly by genotype, we expect to find that no one is prepared to sacrifice his life for any single person but that everyone will sacrifice it when he can thereby save more than two brothers, or four half-brothers, or eight first cousins. (Hamilton 1964: 16)

The "superorganism" solution and the kin-selection solution are in many ways extremes, and understanding what is tied up with each illuminates foci that are relevant to understanding the task set by anyone who seeks an explanation of social insect behavior in terms of Evo Devo.

The notion that colonies of social insects form an organism entered the modern discussion in 1911, with the publication of W.M. Wheeler's "The ant-colony as an organism." The argument in that paper is that several entities at various levels of organization meet the definition of "organism" and can meaningfully be theorized as such, though we do not usually speak of species or of human societies in that way. Defining "organism" is tricky, of course, and much about Wheeler's argument depends on what he takes organisms to be. In addition to a conceptual thesis, however, Wheeler also articulates a research program:

> If it be granted that the ant-colony and those of the other social insects are organisms, we are still confronted with the formidable question as to what regulates the anticipatory coordination, or synergy of the colonial personnel and determines its unity and individualized course.
> (Wheeler 1911: 320)

For Wheeler, the organism's "individualized course" is directed toward nutrition, "producing other similar systems, known as offspring," and protection from the environment (p. 308).

Wheeler's original articulation of what is now usually called the superorganism concept is worth revisiting here because of the differences between what he thought the appropriate research project might look like in 1911, and what it has become. Wheeler understands colonies to be organisms, and while I will argue in the next subsection that this is not quite right, the organismal approach moves him to include the mechanisms for ontogeny (p. 324) at the colony level in his sketch of work to be done. Wheeler was, in an interesting sense, a developmental biologist.

Against the backdrop of Wheeler's writing, recent discussions of the evolution of behaviors associated with colonial organization have a decidedly neo-Darwinian cast, as much of the conversation is about whether kin selection or some kind of group selection best accounts for them. This has been the case since part of the biological community responded negatively to the strong group-selection arguments in Wynne-Edwards's *Animal Dispersion in Relation to Social Behaviour* (1962), with objections drawn from theoretical population genetics (see, for instance, Maynard Smith 1964, 1974). W.D. Hamilton's famous 1964 papers in the *Journal of Theoretical Biology*, and his 1967 paper in *Science*, moved the discussion further in the direction of a genetic approach, as did G.C. Williams's *Adaptation and Natural Selection* (1966). There have since been a slew of papers and books written from this perspective.

In contrast to Wheeler's project, on which the colony's behavior and its evolutionary history are best understood in terms of (super)organismal evolution and development, most of the more recent work is cast in terms of gene frequency changes tied to kin-group fitness of populations of individual organisms in a larger colony context. There is little attention in this work to development of the organism, let alone to Wheeler's idea that colonies can meaningfully be said to develop. In these studies, explanations are usually given at the level of individual organisms and below, as acts of cooperation or altruism are accounted for by applying versions of Hamilton's rule. It is worth noting that not much has changed in this literature since the theoretical advances of Hamilton, Maynard Smith, and Williams in the 1960s and 1970s.

Levels of development

The Evo Devo approach has the potential to reshape research on the evolution of social insects by integrating a conceptual framework that Wheeler might find familiar with empirical and theoretical work on natural selection and behavioral ecology. In order to do so, however, it will first be necessary to be clear about what the object of study is.

In a recent paper (Hamilton *et al.* in press), my co-workers and I argued that, contra Wheeler, colonies are not properly understood as organisms because the causal relationships that hold between organisms and their parts are not the same as those that hold between other biological entities and their parts: colonies seem not to be rendered cohesive in the same way or by the same processes as organisms. The cohesion-generating relationships that hold between organisms and those that hold between colonies are not the same (Haber and Hamilton 2005; Hamilton and Haber 2006). This is to deny that colonies are organisms, but it is not to say that Wheeler was on the wrong track. We argue that colonies and organisms are not the same thing, but that they are instances of the same kind of thing. Namely, they are both evolutionary and developmental individuals. Wheeler had the right idea, but did not quite get the ontology right.

The concern with individuality and the cohesion-generating relations that partly constitute it may seem like philosophers' abstractions, but understanding the ontology of the system points in interesting directions for the Evo Devo approach. Indeed, this framing of the object of study points back to Wheeler's research program. Merely noticing that we have an individual in some metaphysical sense does not give any advice about what causal processes are relevant, or what the "glue" is that holds the colonial organisms together to form a unified whole, from either an evolutionary or a developmental perspective. We are left, then, essentially, with Wheeler's "formidable question" quoted several paragraphs back: what regulates the individualizing features of the colony?

From an evolutionary perspective, this question can be understood in terms of the debate sketched in the last section: evolution "sees" colonies as units of selection and colonies reproduce differentially. More tightly integrated colonies are fitter, so cooperative behaviors are selected for. This sort of answer drives us right back to a set of process and function questions about how such colonies could have arisen by natural selection, and what role group or kin selection might have played, as well as more complicated questions about what roles (and strengths) simultaneous selection at various levels may have played. In other words, when the ontological framing is tightened, the debate about evolutionary accounts of the rise and maintenance of cooperative behavior shifts but need not change in a way that dramatically reorients research into the topic. These questions can be addressed in the standard way, which is to try to get experimental purchase on the questions and to build models of evolution that make sense of the data (Kern Reeve and Hölldobler 2007; Page *et al.* 2006).

The ontological shift makes a larger difference from the developmental perspective, where it is not all business as usual. If social insect colonies are a kind of biological entity, but they are not organisms, what is there to be said about their development? Even if one grants that they do develop in a properly biological (rather than analogical) sense (Hamilton *et al.* in press; Michener 1974), there is still the important question of what the relationship is between developmental processes at the organism level and the colony level.

The integration of evolution and development that Laubichler envisions for Evo Devo is driven in large part by understanding the mechanisms of development from the gene up and from the organism down. One challenge of using this framework to understand colony form and function is that while gene regulatory networks are sometimes highly conserved across taxa, they are presumably not conserved across levels of organization. What is true of the organism is not obviously true of the colony, even when colonies are understood to be "superorganisms" of some kind.

There is conceptual and empirical work to be done, then, in articulating and testing concrete hypotheses about the development of colonies. Put another way, an Evo Devo whose integrative force comes from understanding the developmental mechanisms for phenotypic variation has a "levels" problem exactly akin to the levels problem in evolution. What we need, at least in the case of social insects, is to understand how the structured complexity that is coming into focus at the level of the gene regulatory network ramifies up through levels of organization to allow a mechanistic explanation for the development and evolution of very complicated phenotypes like cooperative behavior. Whether or not gene regulatory networks scale up is an empirical question, but one motivated by understanding levels, individuals, and cohesion-generating relationships.

CONCLUSION, AND NEW DIRECTIONS

This chapter contains an argument and a challenge. The argument, which draws on the work of William Bechtel and others, is that mechanistic explanations are by their nature reductionistic, but that we should not be worried about the recent focus on genes in Evo Devo because mechanistic explanations are also inherently inter-level. With mechanisms, function is not the only focus: form matters because the organization of the parts is at least as important as the parts themselves.

The challenge is to understand and explain colony-level development, its evolution, and its relationship to organismal development. This is a hard case because so much of the evolutionary story remains unsettled, and because the cohesive "forces" that hold colonies together are less tractable visually and experimentally than those that hold organisms together. Yet if colonies are to be understood as cohesive individuals that evolve, it will be necessary to understand what it means to speak of colony-level development mechanistically. Here I have offered two conceptual tools – cohesion-generating relationships and levels of development – to help frame the discussion of a mechanistic Evo Devo that can throw light on form and function, as well as on the relationships between large-scale individuals and their organismal parts.

Future directions

This chapter happens to stand at the end of the volume, and so a conclusion to the larger work is in order as well. I have chosen to do this by way of setting up a few unsettled issues that I take to be foundational because they point to the necessary conceptual scaffolding for a set of projects centered around evolution and development, rather than to particular questions about how form and function are related. Indeed, I am listing here a set of issues that my own interdisciplinary research group is grappling with.

First, with integration comes complexity, but it is not clear at all what it means to have a complex mechanistic explanation that captures more than the organization of very local phenomena, particularly in ways that integrate causes at different scales. One reason that the shape of such explanations remains unknown is that the philosophical literature on mechanisms is mostly about how to explain particular, relatively simple phenomena in the neurosciences and in cell biology. Even the more general work that has been done on the mechanistic approach as a strategy for explanation and integration is mostly historical and retrospective, and it is difficult to see how one might apply it to the current project of marrying disciplines as far flung as evolutionary biology and developmental biology. This is not to express doubt about making such an approach work. It is, rather, to point out that there is a philosophical literature to be developed that would be of immediate use in helping to move Evo Devo toward its goals.

Second, there is the question of how one might best use the mechanistic approach to articulate an account of the development and evolution of complex phenotypes that is broad enough to include the work that has been done by those who have employed other explanatory strategies. Given that much of the background is conceptual and that some of the important tools available – models of population dynamics and mathematical descriptions of morphospace, for instance – might better be called theoretical than mechanical, there is work to be done on how to integrate mechanistic explanations with theoretical and computational tools of various kinds in a way that is both illuminating and remains true to the mechanistic emphasis on limning the causal structure of the system of interest.

Finally, talk of levels-of-development and other discussions of major and minor biological transitions in this volume lead back to the old problem of emergence. This is a topic on which very little of use to biologists has been written, despite the fact that so many of the ongoing questions in biology have to do with complicated, or even complex, relationships between levels of organization. This situation has come about partly because many of the researchers who work on emergence have not done so as part of an active biological research program. This volume provides a great deal of ground – in terms of both information and case studies – for basic spadework for anyone who wants to roll up his or her sleeves. Providing such ground for new discussions on this and many other topics has been a driving concern behind this collection.

REFERENCES

Bechtel, W. and Abrahamsen, A. (2005). Explanation: a mechanist alternative. *Studies in History and Philosophy of Biological and Biomedical Sciences* 36, 421–41.

Bechtel, W. and Hamilton, A. (2007). Natural, behavioral, social sciences and the humanities: reductionism and the unity of the sciences. In T. Kuipers (ed.), *General Philosophy of Science: Focal Issues*. Amsterdam: North Holland, pp. 377–430.

Bechtel, W. and Richardson, R.C. (1993). *Discovering Complexity: Decomposition and Localization as Strategies in Scientific Research*. New Jersey: Princeton University Press.

Darwin, C.R. (1871). *The Descent of Man, and Selection in Relation to Sex*. New York: D. Appleton and Company.

Davidson, E.H. and Erwin, D.H. (2006). Gene regulatory networks and the evolution of animal body plans. *Science* 311, 796–800.

Glennan, S. (2002). Rethinking mechanistic explanation. *Philosophy of Science* 69, S342–S353.

Haber, M. and Hamilton, A. (2005). Coherence, consistency, and cohesion: clade selection in Okasha and beyond. *Philosophy of Science* 72, 1026–40.

Hamilton, A. and Haber, M. (2006). Clades are reproducers. *Biological Theory* 1, 381–91.

Hamilton, A., Smith, N.R., and Haber, M.H. (in press). Social insects and the individuality thesis: cohesion and the colony as a selectable individual. In J. Gadau and J. Fewell (eds.), *Organization of Insect Societies: From Genome to Sociocomplexity*. Cambridge, MA: Harvard University Press.

Hamilton, W.D. (1964). The genetic evolution of social behavior. I and II. *Journal of Theoretical Biology* 7, 1–16, 17–32.

(1967). Extraordinary sex ratios. *Science* 156, 477–88.

Kern Reeve, H. and Hölldobler, B. (2007). The emergence of a superorganism through intergroup competition. *Proceedings of the National Academy of Sciences USA* 23, 9736–40.

Lewontin, R.C. (2000). The dream of the human genome. In R.C. Lewontin, *It Ain't Necessarily So: The Dream of the Human Genome Project and Other Illusions*. New York: Basic Books, pp. 135–76.

Machamer, P.K., Darden, L., and Craver, C.F. (2000). Thinking about mechanisms. *Philosophy of Science* 67, 1–25.

Maynard Smith, J. (1964). Group selection and kin selection. *Nature* 201, 145–7.

(1974). Group selection. *Quarterly Review of Biology* 51, 277–83.

Mayr, E. (1961). Cause and effect in biology. *Science* 134, 1501–6.

Michener, C.D. (1974). *The Social Behavior of the Bees*. Cambridge, MA: Belknap Press of Harvard University Press.

Minelli, A. (2003). *The Development of Animal Form: Ontogeny, Morphology, and Evolution*. Cambridge University Press.

Page, R.E., Scheiner, R., Erber, J., and Amdam, G. (2006). The development and evolution of division of labor and foraging behavior in a social insect (*Apis mellifera L.*). *Current Topics in Developmental Biology* 74, 253–86.

Raff, R. and Sly, B. (2000). Modularity and dissociation in the evolution of gene expression territories in development. *Evolution & Development* 2, 102–13.

Wagner. G.P. (1996). Homologues, natural kinds and the evolution of modularity. *American Zoologist* 36, 36–43.

Wheeler, W.M. (1911). The ant-colony as an organism. *Journal of Morphology* 22, 307–25.

Williams, G.C. (1966). *Adaptation and Natural Selection*. New Jersey: Princeton University Press.

Wilson, D.S. and Sober, E. (1989). Reviving the Superorganism. *Journal of Theoretical Biology* 136, 337–56.

Wimsatt, W.C. (1974). Complexity and organization. *PSA: Proceedings of the Biennial Meeting of the Philosophy of Science Association* (1972), 67–86.

(1976). Reductive explanation: a functional account. *PSA: Proceedings of the Biennial Meeting of the Philosophy of Science Association* (1974), 671–710.

Wynne-Edwards, V.C. (1962). *Animal Dispersion in Relation to Social Behaviour*. New York: Hafner.

Index

www.ingramcontent.com/pod-product-compliance
Ingram Content Group UK Ltd.
Pitfield, Milton Keynes, MK11 3LW, UK
UKHW040708240125
453752UK00026BA/13